T0271249

Teaching Children to Love Problem Solving

A Reference from Birth through Adulthood

Problem Solving in Mathematics and Beyond

Series Editor: Dr. Alfred S. Posamentier
Chief Liaison for International Academic Affairs
Professor Emeritus of Mathematics Education
CCNY - City University of New York

Long Island University
1 University Plaza -- M101
Brooklyn, New York 11201

Published

**Problem Solving in
Mathematics and Beyond** | Volume **06**

Teaching Children to Love Problem Solving

A Reference from Birth through Adulthood

Terri Germain-Williams

Mercy College, USA

World Scientific

NEW JERSEY · LONDON · SINGAPORE · BEIJING · SHANGHAI · HONG KONG · TAIPEI · CHENNAI · TOKYO

Published by

World Scientific Publishing Co. Pte. Ltd.

5 Toh Tuck Link, Singapore 596224

USA office: 27 Warren Street, Suite 401-402, Hackensack, NJ 07601

UK office: 57 Shelton Street, Covent Garden, London WC2H 9HE

British Library Cataloguing-in-Publication Data
A catalogue record for this book is available from the British Library.

Problem Solving in Mathematics and Beyond — Vol. 6
TEACHING CHILDREN TO LOVE PROBLEM SOLVING
A Reference from Birth through Adulthood

Copyright © 2017 by World Scientific Publishing Co. Pte. Ltd.

All rights reserved. This book, or parts thereof, may not be reproduced in any form or by any means, electronic or mechanical, including photocopying, recording or any information storage and retrieval system now known or to be invented, without written permission from the publisher.

For photocopying of material in this volume, please pay a copying fee through the Copyright Clearance Center, Inc., 222 Rosewood Drive, Danvers, MA 01923, USA. In this case permission to photocopy is not required from the publisher.

ISBN 978-981-3209-82-4
ISBN 978-981-3208-79-7 (pbk)

Desk Editor: Tan Rok Ting

Printed in Singapore

For my boys, EJ, Zekie, and Eli.
Mommy loves you
and appreciates you
for working on your coloring and puzzles alongside me,
patiently waiting for me to come play.

Preface & Acknowledgements

This book was developed with the caring and concerned adult in mind. This text is a one-stop for anyone who would like to help a child develop problem solving thinking and become adept at the use of problem solving strategies.

I wish to express my gratitude to those who took the time to provide thoughtful feedback and guide this work. My deepest appreciation to Alfred Posamentier for the opportunity to join as an author of this prestigious series. For your patience with me and for always helping me becoming better, I will be forever grateful.

Incalculable thanks to Devin Thornburg, all your notes, ideas, and the impact of everything you've gained in your thoughtful career are reflected in these pages. Your wisdom and guidance are so very appreciated.

Mary Lambert, my student teaching supervisor, who was gracious enough to provide a keen literate eye for typos and omissions, but also share the perspective of the parent without a mathematics background. You will forever hold a very special place in my heart, from the moment I walked into the UC Ballroom at Adelphi University and scanned the room, hoping that it was you who was assigned my clinical supervisor.

For my friends, Rose Keeler-Ahern and Daniel Vessely, who not only took the time out of your busy lives to read this manuscript and provide feedback from your perspective but who also stood by me while I wrote so very many hours that I might have been enjoying your company if not otherwise engrossed in the work of this book.

Introduction

"If I had an hour to solve a problem and my life depended on the solution, I would spend the first 55 minutes determining the proper question to ask, for once I know the proper question, I could solve the problem in less than five minutes." – Albert Einstein

Problem solving is a critical skill for success in any field or endeavor. The path to success on a project or goal is bound include pitfalls and challenges. The ability to persevere when faced with difficulty or the unknown will impact the outcome of any trajectory. Problem solving skills are not natural but they can be developed through practice and modeling.

George Polya, leading thinker in problem solving published his work, "How to Solve It" that outlined the following four-step process for addressing problems:

(1) Understand the problem
(2) Devise a plan
(3) Carry out the plan
(4) Examine the result

In this book, problem solving is addressed within the context of the stages of development. Anyone with a child in his or her life can use this reference to understand the range of a child's potential for each age. There are problems that can be assigned or worked through together to develop a child's thinking. The recommended activities can become part of the weekly routine to regularly engage in practices and conversations to develop problem solving naturally.

Each chapter begins by describing the child's cognitive ability at this age range. Next, an overview of key mathematical ideas that students might be learning in an educational setting is included. In the third section of each chapter, sample problems are posed with solutions and methods for solving. In the final section, suggested games, activities, challenges, conversations, and contextual learning are included to ensure that techniques are being developed continuously and at times outside of the pencil-and-paper setting.

The *manner* in which a problem is approached is extremely influential in the outcome, as you may have also found in situations you've encountered. Throughout your interactions with children, make every effort to model and encourage the characteristics of positive mindset and perseverance. Children learn much from the ways in which we, the adults in their lives, approach problems — particularly when the problems are difficult.

In addition to mindset, while not exhaustive, the following is a list of Problem Solving Strategies that might be used to solve new problems.

(a) Act it Out
(b) Analyze the Units
(c) Convert to Algebra
(d) Create a Physical Representation
(e) Use Deductive Logic
(f) Draw a Diagram
(g) Draw Venn Diagrams
(h) Eliminate Possibilities
(i) Evaluate Finite Differences
(j) Guess and Check
(k) Identify Sub-problems
(l) Look for a Pattern
(m) Make a Systematic List
(n) Produce a Model
(o) Organize Information
(p) Solve an Easier Related Problem
(q) Use Matrix Logic
(r) Visualize Spatial Relationships
(s) Work Backwards

Many of the problems in this book are solved using multiple strategies. You may enjoy the challenge of finding a solution using a strategy other than those included in this text!

A Note About Incorrect Answers

"We must help the child to act for himself, will for himself, think for himself; this is the art of those who aspire to serve the spirit," wrote Dr. Maria Montessori.

There will be moments that arise where the child proposes an incorrect answer. The child may be emphatic or insistent! Our instinct may be to correct immediately or identify incorrect thinking. Research has demonstrated that responding in a positive way to incorrect answers can be integral to developing resilience and promoting continued motivation. These practices are applicable to children of every age. You might begin by reflecting back what you have heard:

- Do you're saying…
- What I think you concluded was…
- I see, so, you think that…

Hearing you repeat back the answer might insight additional thinking or the discovery of an error. Positive responses can also be crafted in the form of questions. You might ask:

- What makes you think this?
- Tell me about a similar problem you've encountered.
- What makes you say…?
- Illustrate your thinking.

Prodding for justification can promote additional positive dialogue about the problem:
- Can you justify your answer?
- Can you provide additional examples?
- Show me.
- Explain your thinking.
- What would you say to someone who disagreed with this?

In addition to the examples in this book, you might consider modeling resilience and motivation by explicitly sharing with a child your own experience. You might tell them, "Well I can think of five reasons not to, but I am going to go ahead with this task." You also might share your process when you have overcome an obstacle. What did you do when you encountered a difficult task? How did you process a review or school grade that you were unhappy with? Troubleshoot when lessons come up in life.

Finally, if you come across a term you are unfamiliar with or a concept you were not taught, please do not hesitate to research any of the topics in this text alongside them. You will be modeling perseverance. You will find many open source (read: free of charge) resources including videos and examples and definitions at the library and on the web. You may find that your child makes connections and then explains ideas to you. And that is one of the most powerful experiences a child can have. The amount of confidence instilled in a child who can learn and teach an idea (especially to an adult) is invaluable!

"Never help a child with a task at which he feels he can succeed."
– Maria Montessori.

Contents

Chapter 1

The Earliest Years: Babies into Infancy

1.1 Infant Development

A new baby brings about so much change in the lives of loved ones. There are many needs of the baby to be met. While many texts address the truly important aspects of physical health and sleep, feeding and changing, this book will start with a baby's understanding of the world as it relates to logical and mathematical thinking that will eventually be of great use in problem solving.

During the first year, of which the first nine months have been sometimes called the "fourth trimester," babies have a distinct need for physical touch and the general feeling of being loved and cared for. Psychosocial theorist Erik Erikson purports that the development of trust during this period of an infant's life is of primary importance. There are a multitude of developmental experts who agree that babies who have been cared for in this vulnerable time are better off in their later years.

1.2 Mathematics for Babies

The thought of a little newborn with mathematical skills may seem far-fetched. Humans are born with very few instinctual skills, which include rooting and sucking, simple motor skills (like the patellar knee response and blinking), and of course crying. Their development in the first year happens quickly and within a few months, babies have developed the fundamentals for some of their much-needed later abilities. Babies track objects as they move, for example, and hold gaze while examining objects.

The tendency to explore objects with our mouths and gums thankfully wanes as they grow.

Even as early as three months of age, some babies have been known to begin to categorize. By one year, grouping of objects by characteristics such as soft and round can occur. There have been studies that also indicate that babies as young as six months old can differentiate between sets of items when the ratio is sufficiently great. Specifically, a baby might be more interested in a pile of 15 grapes versus a pile of 7 grapes but not have a preference when the piles are closer in number, such as 7 grapes versus 10 grapes.

Babies are problem-solvers as they investigate the cause-and-effect of their actions, especially after the first few months of life. They are listening and beginning to understand the idea of "more," particularly in their own context of food or drink. You might see your baby turn items over to discover whether an item's front matches its' back. As the months pass, you will notice your growing little one throwing food or banging objects and then being very interested in your reaction to what s/he just did.

1.3 Sample Problems for Baby

Author's Note: In subsequent chapters, this section will provide problems with worked-out answers for the age group in the chapter. Since babies are not quite ready to embark upon problems, there are no problems in this chapter. Increase time on the tasks.

1.4 Everyday-Problem-Solving-Baby

Reading to your baby as much as possible has shown to be vitally important to cognitive development. Language learning is taking place. Babies will begin to respond to their name. They will begin to mutter sounds and develop associations with key words. Their vocabulary will remain limited within the first year, but increased exposure to words and association with letters or pictures has been associated with higher reading levels and increased vocabulary in later years.

Although there is a plethora of items available for babies, the items a baby needs need not break a budget. As baby's sight develops, a mobile placed above the crib (which can be homemade, as recommended by childcare author Burton L. White). Accessible baby mirrors are recommended for babies to view movement. Babies should also have age-appropriate, clean objects to hold on to and chew as they explore their surroundings. Balls and teething rings and infant books can be very exciting for babies to see! Arm movements are developed during this period, as you will find they knock over much of anything within their reach. There are also mats equipped with hanging items for babies to fixate and tap with their feet and hands. Floor pillows also come with attached toys for very young babies to practice laying on their bellies since they are put to sleep on their backs. Babies should also be exposed to elements of their environments through taking them for a walk whether it be indoors or outdoors. In the later months of the first year, children become especially interested in hinges that allow them to open and close objects. Infants have a distinct interest in opening and closing drawers and cabinets within their reach.

Vocal communication during this time is very important, as infants are developing their knowledge and associating what you are saying with what they are seeing or doing. Identifying objects and actions will help the baby continue to associate with their correct terminology. Research has indicated that while videos or books may sometimes supplement a child's vocabulary, hearing words from humans surrounding them has the greatest impact on learning speech.

Chapter 2

Developing Toddler Thinking

2.1 Toddler Development

You will notice some major developments after your child's first birthday. Encouraging the scientist in your little one will bring about explorations and learnings at a pace that will be difficult to keep up with. They are becoming imitators and will sing lyrics or repeat quotes that you may not even recall them having been exposed to. During this time, there are many ways to ensure that they continue to remain inquisitive and begin to understand cause and effect.

By this time, toddlers have now acquired the ability to obtain and sustain the attention of caregivers. They can ask for help when they are unable to complete a task on their own. They are also beginning to take on imaginary roles such as mom or dad or a storybook character. They are also beginning to act out situations and pretend to engage in helping. At the same time, the understanding of discipline and consequences can lead to some negative feelings and interactions.

2.2 Toddler Mathematics

Consider opportunities to make numbers part of your everyday routine. When you are engaging with a repeated activity, use this time as an opportunity to practice counting, such as, when walking down stairs together, count each step. The repetition of counting will be reinforced and you'll have a child who can count in no time! When reading a book with multiples of the same items, pause to count, for example, "How many birds in this photo?" Count the steps in getting dressed, "One arm, two arms."

"One Head." This dialogue will also come in handy if you have a child who resists wardrobe changes: counting out the steps will be a familiar routine, and a resistant child may count with you to pass the time and get through the transition more smoothly.

Puzzles can also be opportunities to teach fundamental vocabulary such as: shapes, colors, letters, animals, parts of items, etc. You might also keep puzzles, for this age, in Ziploc bags to prevent missing pieces and reinforce clean-up/put away of each puzzle before choosing the next one.

2.3 Everyday Ideas to Support Problem Solving

To connect with section 1.3, toddlers are still a bit young to be solving specific, mathematics-related problems. Still, there are many ways you can support their development. For one, when climbing or trying new steps, stay close to your toddler to help if necessary, but you would be surprised how their balance and kinesthetic sense will develop when provided with the opportunity to try new ways to use their hands and feet. Allowing them to climb on play-spaces, slides, chairs, stairs (with you in arm's reach), and low beams will be integral to their understanding of how their bodies can be used to balance and to support themselves.

Make every attempt to allow toddlers to discover their world and to minimize or eliminate screen time. The American Academy of Pediatrics (2001) published the following recommendation:

> *Television and other entertainment media should be*
> *avoided for infants and children under age 2. A*
> *child's brain develops rapidly during these first*
> *years, and young children learn best by interacting*
> *with people, not screens.*

If and when you decide to introduce screen time, carefully preview the material, as even programs that may be promoted for your child's age, may not be demonstrating the problem-solving skills you are wishing to impart and may, to the contrary, be promoting unhealthy habits. Consider reviewing a prospective show by watching it and accounting for the number of instances of, for example, kindness, violence, bullying, product placement, positive interactions with school and parents, materialism, generosity, and use of manners. Allow children to play outdoors as much as possible.

The emphasis on reading cannot be reiterated enough, as children are now very interested in being read to in animated and engaging short stories. While their attention may not be held for long, they are very interested in attempting to memorize or understand the pages of books. Meta-analysis study found better writing in preschoolers who had been read to, regardless of socioeconomic class (Bus and van IJezendoorn, 1995).

2.4 Toddler Problem Solving Activities

While most toddlers are in the very early stages of verbal development which prohibits the use of many types of "traditional" written or oral problems, there are many ways to address the development of critical thinking in this age range.

Activity 2.1 Making the Most of Memory

As your child begins to make sense of time, begin alerting them of activities to come. For example, you might tell your toddler, "After your nap, we are going to take a trip to the park." And when your toddler awakens, you might ask, "Do you remember where we are going now that you have taken your nap?" This would be more appropriate for a one-year-old. As your child approaches 2 years of age, begin to talk about activities for "tomorrow" and begin to incorporate the days of the week. You might even demonstrate the days using a calendar, "Tomorrow is Saturday. We are visiting Grandma tomorrow."

Provide your child the opportunity to make decisions and choose. They will activate memory about an object or action, and feel empowered in ways that will help in the future when they begin to test their environment and choice-making. You might ask, "Would you like to wear the blue pajamas or the red pajamas?" Also, keying into memory will also be important for daily functions, such as waiting for foods to be cut before eating and blowing on food that is "hot." Toddlers begin to imitate actions they see you doing such as opening and closing doors, turning faucets on/off and flushing the toilets. Teach them moderation in these activities, such as to gently close doors, count to three and turn off the faucet to conserve water.

Activity 2.2 Spills & Messes are opportunities to emphasize Cause & Effect

Do your best to remain calm and encouraging when your toddler turns off the surge protector that knocks out your internet, empties your cabinets, spreads dirt onto your walkway (or him/herself), wipes the peanut butter off the apple and spreads it all over his/her face, and pours out his/her drink. Toddlers are not attempting to "make a mess" nor are they looking to delay your departure or make you upset. They are exploring both their environment and your reaction at the same time. In lieu of making punishment and admonishment for normal messes, work with your toddler to understand the actual consequence (effect) of the behavior. Here are some examples:

Table 2.4-1: Cause and Effect

Cause	*Effect*
Mess!	**React!**
Pour out cup of liquid	Demonstrate wiping the mess with a towel and provide your toddler with a cloth or share your cloth to help clean it up.
Empty a cabinet of pots and pans	Allow your child to explore and bang a bit. When s/he is ready to move on to the next cabinet or toy, sit and help your toddler clean up all of the items and put them back into place.
Transporting or pouring dirt onto table or another inappropriate place	Provide your toddler with a small broom to help wipe the dirt back to where it belongs
Messy or sticky substance on hands or face or body or toy	Provide a wash cloth for toddler to begin to wipe and clean themselves or the dirtied object

The above-mentioned messes and tests were not inherently dangerous. In the case of dangerous behavior (such as biting an electrical cord, emptying the knives from a drawer, playing with stove/oven buttons, pulling on the cord from a window blind, emptying the cupboard of cleaning supplies, or playing with electrical outlets), identify the location or behavior as dangerous and make clear that you do not wish your child to get hurt. Clearly discuss behaviors and locations that are not acceptable. You will be surprised how much your little one understands and will follow your concerned directions. This will also help your toddler problem solve when approaching activities: sometimes it's ok to play with a string (such as when attached to a toy) but sometimes it is not (such as when it is attached to an appliance or window treatment).

Tip: Consider singing a clean-up song with your toddler when cleaning up. This will allow your toddler to have some amusement cleaning with you, and associate the tune with a positive view of putting messes away.

Activity 2.3 Repetition: Challenging for us, Productive for them

Toddlers will ask for the same things over and over, and will also perform actions over and over again. This may include a song they wish to hear or a show they wish to watch or stairs they wish to climb or a button they wish to continually press. The key for us is to provide them with safe and productive opportunities to repeat. They might use building blocks to build and then knock down. They might connect two blocks only to rip them apart repeatedly. They might push buttons on a computer, phone or remote control. When possible, provide them with an age-appropriate replica.

Let them repeat as much as possible, and also take good care to choose songs and activities wisely. Provide age appropriate songs and CDs for them to engage with.

Activity 2.4 Provide Support and Minimal Intervention

When your child attempts a puzzle, refrain from turning the piece or putting it together in haste. Let your child explore, even if your child needs to take the time or try it again another time. Puzzles are a wonderful opportunity for children to explore problem solving. You might suggest "What if you turn it?" which will teach them the word "turn" while allowing them the opportunity to learn patience and multiple attempts. Pay attention to your child's abilities. You might find that the puzzles quickly become quite easy and they can be challenged to do more difficult puzzles. Be sure to continually provide the next level of challenge.

Activity 2.5 Develop Puzzle Skills

Try encouraging the exploration of puzzles by sorting pieces. They might be sorted by color or size in non-connecting puzzles, then by the number of straight edges for interlocking puzzles. Trace the pieces of the puzzle. Identify characteristics of the puzzle piece. You might also have your child trace each piece as its place is found in the puzzle. Finally, consider creating puzzles from coloring books, photos, newspaper, or magazine pages.

Table 2.4-2: Puzzle Progressions

Non-Connecting Puzzles with Large Knobs

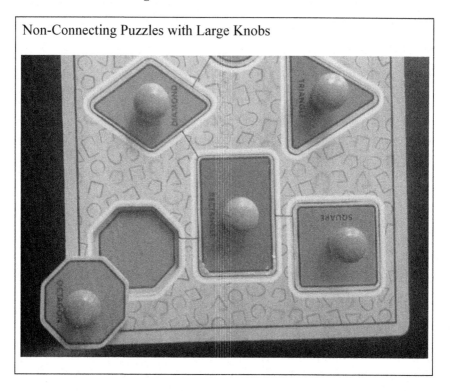

Non-Connecting Peg Puzzles

Inset Non-Connecting

Foam Inset Non-Connecting Puzzles

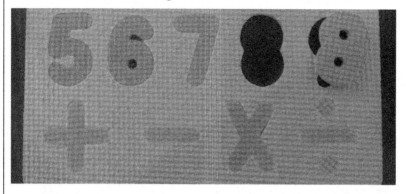

Mosaic Inset

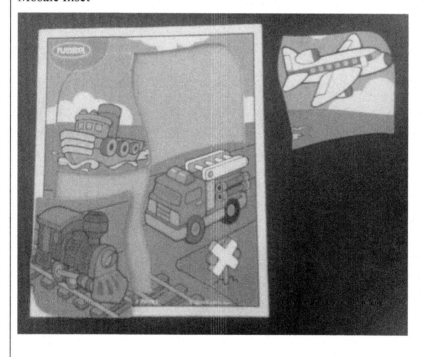

Complex Inset Puzzle

Inset   Interlocking

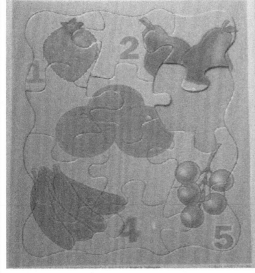 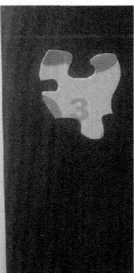

Interlocking Puzzles

Chapter 3

Preschool and Problem Solving

3.1 Development of Preschoolers

Preschoolers' language is exploding and they are more than excited about exchanging information concerning just about anything during this time. They are enjoying creating personas and scenarios during imaginative play. They are also developing logical reasoning. They can now sort objects and determine the commonality that brought each set of items together. They are inquisitive about their environment and enthusiastic about accomplishments and new information. Their brains are extremely busy growing at a rapid pace, building neurons and creating synapses.

This is also the beginning of some formalized education for some students, where they will be guided by an instructor in a class or group setting. It is also at this age where the national organizations in the United States begin to outline specific curriculum.

While their attention spans may seem short at a length of about five minutes, they have increased significantly since birth. In addition to a short attention span, another drawback of this period is the lack of ability to understand another person's perspective.

3.2 Mathematics for Preschoolers

Counting is one of the most important skills to begin with your child in these formative years. At this age, children should count orally to twenty in correct sequence. They should be able to count items in groups up to 10. Many children at this age are also beginning to trace, copy, and practice writing numbers.

Children will be identifying and categorizing shapes in both two and three dimensions. They can apply the shapes to patterns, drawing, and can describe them using some of their attributes.

At this age, measurement concepts are emerging. Equal lengths can be determined by the preschooler. Lengths can be compared using the words "longer" or "shorter." Children are also beginning to do elementary data analysis as they might compare the number of items they have to the number of items a peer or sibling has. They can use tally marks to take record of objects or events. A chart in a classroom might tally how many rainy versus sunny days have occurred during a given time frame. They might count how many blue building blocks are in a box versus how many red or yellow blocks are included.

Table 3.2-1: Preschool Terminology

Practice identifying and defining the following terms with your preschooler:		
0, 1, 2, 3, 4, 5	How many	short
behind	in front of	small
big	last	sort
bottom	light	sphere
circle	next to	square
down	over	tall
empty	on	top
first	pattern	triangle
full	rectangle	under
heavy	shape	up

3.3 Sample Problems for Preschoolers

Problem 3.1

"We have 4 chairs at the table. If we take away one chair, how many people can sit at the table?"

Possible Solutions

Strategy: Act It Out

The child might move a chair away from the table and count the number of chairs. The child might also refrain from counting one of the chairs in lieu of removing it from the table.

Strategy: Create a Physical Representation

The child might have chairs or seats from a play set that they might grab to represent the large chairs.

Strategy: Produce A Model

The child might use a number line or other set of sequential numbers to solve the problem. A number puzzle (where the numbers can be removed) might look like this:

Figure 3.3-1: Number Puzzle

The child could begin with the 4 and count backwards one to find 3.

Strategy: Draw a Diagram

The child might use a pencil and paper to draw the table and mark chairs, crossing one out and counting the remainder.

Problem 3.2

While at the store, you might use the collections of objects for purchase as an activity. "There are three apples in the produce bag. If we add two, how many will there be in the bag?"

Possible Solutions

Strategy: Act It Out

The child might count out two more apples and put them into the bag, and then count the new total. The child may also point to three apples and count the new apples first, continuing with the ones in the bag.

Strategy: Draw a Diagram

The child might use a pencil and paper to draw three apples and then draw two more, then count the total.

Strategy: Produce A Model

The child might use their fingers on a number line or other set of sequential numbers in a puzzle or poster or game to solve the problem. The child could count three first, and move two to the right to find the total of five.

Problem 3.3

What items in this room are as long or as wide as this book?

Possible Solutions

Strategy: Create a Physical Representation

The child could explore item and bring them over to compare to the length of this book. The child might pick up a writing utensil and compare it to the book. S/he might designate the item "too short." Another book in the room might be "too long."

Strategy: Analyze the Units

The child could use a ruler, paper, arm, or string to create a unit of measure the same length as the book. The child could take the unit of one book length around the room, finding matches!

Strategy: Create a Model

The child could trace the book on a piece of blank paper to bring around the room, comparing items to the size of the outline of the book.

Problem 3.4

Continue this pattern:

Strategy: Draw a Diagram

The child might use a pencil and paper to recreate the pattern and then continue it. Notice this question is open-ended so you can follow the child's lead in how long the pattern will continue.

Strategy: Produce A Model

The child might cut out objects or find similar shapes among cards or toys to continue the pattern.

Problem 3.5

Sort the following objects:

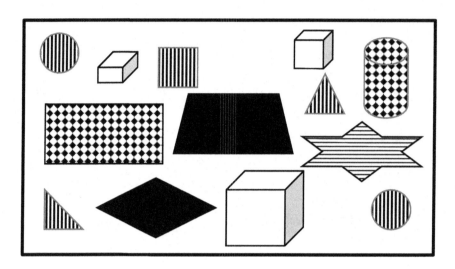

Possible Solutions

In this case, you are allowing the child to determine the possible sorting options. Here we have a number of possibilities for sorting. You might see sorting by shape, by size, by fill, or by dimension. You could add a challenge to this problem by saying: sort into two groups or sort into three groups.

If you have the opportunity to reproduce this activity in your home using toys, writing utensils, blocks or kitchen items, you can create sorting activities that your little one can do hands-on.

Problem 3.6

Choose an item that will create a road that is as long as this truck:

Here are the options:

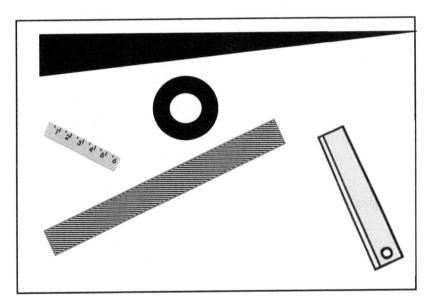

3.4 Everyday Problem Solving for Preschoolers

Activity 3.1 Rhythm for Mathematics

Music can be a wonderful way to create connections between mathematical thinking and activity. Patterns can be demonstrated via rhythms, especially in popular nursery rhymes. Clapping or playing instruments or even tapping can support the connection between movement, rhythms, and pattern-making.

Activity 3.2 Sorting Together

Provide multi-colored building blocks, pattern blocks, money, or objects and practice sorting. Let your child determine the categorization method and also suggest patterns that the child may not think of.

Activity 3.3 Original Patternmaking

Use objects, sounds, or words to create patterns. Ask your child to continue the pattern for a few more objects. Ask your child to create a pattern for you to continue. Identify patterns in the places you go and discuss what you notice. You might find patterns in the bricks of a walkway or floor. You might find patterns in the arrangement of items in a store or at a library or museum. Patterns can frequently be found in fabrics or items of clothing.

Activity 3.4 Shape Discovery

Play games where you seek to find shapes or colors along your journey together. This can be also used as a counting activity. You might ask, "How many squares do you see in this book (or room or pattern)?" or "Let's see how many rectangles we can find on the way to our friend's house."

Activity 3.5 How long is it?

Begin to talk about length. At this time there is not a need for customary units (metric or otherwise) nor rulers. Ask your child, "how many cars long is this mat?" or "how many crayons would fit across the top of this page?" or "how many steps long is this driveway?" Estimating lengths is a critical skill for unitary measure and geometric thinking later in life.

Activity 3.6 Incorporating and Reinforcing Countdowns

Counting down can be incorporated into everyday activities such as mealtime or clean-up time. You can say "I see six cars that need to be put away" or "We are each going to each eat five green beans before being excused." Children can practice counting backwards from the given number of items.

Activity 3.7 Which one is smaller?

Use flash cards, dice, or pencil and paper to compare numbers. You can provide your child with some examples, and then let him or her quiz you.

Activity 3.8 Creating Shapes from Scratch

Provide materials and create the four basic shapes (squares, circles, triangles, rectangles) from material such as paper, sticks, clay, dough, etc.

Activity 3.9 Patternmaking

Challenge your child to create a pattern using colors and shapes for you to guess how it continues.

Chapter 4

Early Childhood Problem Solving

4.1 Development in Early Childhood (5, 6, & 7-Year-Olds)

Around age five, a child's attention span has doubled to about 10 minutes long. Many students, when attending to something they are extremely interested in, can stay engaged for 15 minutes or even 20 minutes. Children are reading an increasing amount of words and sentences, so they are able to engage with mathematics in new ways. They are also becoming more social and less ego-centric as they progress through early childhood. Children are becoming more interested in working together and discussing their progress with their peers.

Gross and fine motor skills are becoming sufficiently refined for activity in a formal classroom such as writing in a workbook or cutting out a design. Many students are also able to jump, climb, and skip at this age, as their coordination and motor skills are supported by their mental and physical development. While not yet fully logical thinkers, children at this age are thinking logically about mathematics. Children can now sort objects into multiple categories and in multiple ways independently. For example: where at one time they were applying concrete attributes, they can now classify items as short, medium, and long. Their recognition of patterns is also becoming more sophisticated. They see patterns in numbers where more than one number is skipped and they can apply mathematical operations to pattern-making.

Memory is greatly improved, and students are ready to add frequently-referenced calculations to their recall memory. These might be sums that add to one hundred that would be helpful in making change for a dollar, for example. They might be doubling numbers. Another example might be subtracting numbers from 60, which is useful in telling time and determining how many minutes left in the hour. Finally, knowing mathematics all addition facts through ten is critical prior to entering third grade. This is important because the ability to quickly recall computations has been shown to increase success in mathematics in later years.

4.2 Mathematics for Early Childhood (5, 6, & 7-Year-Olds)

During these years, children are developing the knowledge of the fundamentals of later mathematics. Children will be able to both count and "skip count," or count by tens, to 120. They will count aloud from a number other than one. They will compare numbers (within ten) either as written numerals or as a number of objects. For numbers greater than 10 they will begin to apply concepts of place value to understand the meaning of double-digit numbers.

At this age, addition and subtraction are introduced. Subtraction has four types of scenarios: take away, comparison, completion, and missing addend (the number being added to another in an addition sentence). Children will be taught to compose and decompose numbers through the use of number bonds (visualization of how numbers can be combined to make larger numbers). They will use visuals and manipulates to determine how many ways 7 counters can be broken down. They are beginning to understand that 7 is counted after 6 because 7 is one more than 6.

They will begin to measure and describe objects by length and weight as well as compare items of different length and weight. They can create shapes from paper or clay. They will be composing shapes from other shapes, as in: put two or four triangles together to make a rectangle using pattern blocks or paper or models.

They will also be able to identify non-defining attributes of shapes (e.g. size, color, orientation).

Figure 4.2-1: Pattern Blocks

Table 4.1: Terminology for Early Childhood

Practice identifying and defining the following terms:		
Orally count 1-100	Fourths	Taller/Shorter
Above	Half / Halves	Third/Thirds
Below	In front of	Three-dimensional
Beside	Money, dollar, cent	Trapezoid
Cone	Next to	Two-dimensional
Cube	Prism	Vertex
Cylinder	Quarters	Vertices
Data point	Rectangle	Write and tell time
Flat/ Solid	Sphere	Write numbers 1-20

By the end of first grade, they will understand the units of time including hours, minutes, and seconds. They should be able to write and tell time using both analog and digital clocks.

4.3 Sample Problems for Early Childhood (5, 6, & 7-Year-Olds)

Problem 4.1

What are three different addition number sentences that conclude with a sum of 4?

Strategy: Create a Physical Representation

The child could work with cards, cut-outs, or objects that represent whole numbers. The objects can be put together and apart to make the total add up to four. For this activity, there would need to be multiple representations of the same number.

Strategy: Produce a Model

The child could use four number cubes, blocks, or objects, splitting them up in many different arrangements to represent groupings within four of the items.

Possible Solutions

Here are the possible number sentences:
$1 + 1 + 1 + 1 = 4$
$1 + 2 + 1 = 4$
$1 + 1 + 2 = 4$
$2 + 1 + 1 = 4$
$2 + 2 = 4$
$1 + 3 = 4$
$3 + 1 = 4$

Problem 4.2
What is the next number in this pattern: 2, 4, 8, 16?

Strategy: Look for a Pattern

What happens between 2 and 4? What happens between 4 and 8? What happens between 8 and 16? The differences can be analyzed for understanding of this pattern.

Possible Solutions

The child may see that the number is added to itself to obtain the next number: $2 + 2 = 4$; $4 + 4 = 8$, $8 + 8 = 16$. Therefore, the next number would be $16 + 16 = 32$

$$2 \ ^{+2} \ 4^{+4} \ 8^{+8}$$

The child may use the term "doubled" to describe the relationship between the number and the next number.

$$2^{x2} \ 4^{x2} \ 8^{x2}$$

Problem 4.3 There were 16 friends on the playground. If four were on the swings, how many were playing elsewhere?

Strategy: Act It Out
This can be acted out in a classroom or social setting, where a total of 16 peers can stand to represent the total number, and four students can be set aside representing those on the swings.

Strategy: Guess and Check

If there were 4 students on the swings, maybe (for example 10 is our guess and) there were 10 playing elsewhere. If this solution isn't correct, analyze the answer: should we have fewer children playing elsewhere or additional children? How will adding or subtracting from the 10 impact our total?

Strategy: Eliminate Possibilities

In this strategy, we can use counting down to find the solution. If there were 16 friends playing elsewhere, there would be zero on the swings. If there were 15 playing elsewhere, there would be one on the swings. If there were 14 playing elsewhere, there would be two on the swings, etc.

Problem 4.4

Six children went to the painting table and two went to the science center. How many more went to the painting table than the science center?

Possible Solutions

Strategy: Act It Out

In this strategy, a child can take six dolls or stuffed animals or even friends to stand together representing the six children. Two people or objects can be moved to a different area representing the science center. The remaining can be counted to determine the result.

Strategy: Draw a Diagram

In this example, a child might draw six circles or figures to represent the six children. Then, two of the objects might be circled or annotated to indicate that they will be moving to the science center, and the remaining counted.

Problem 4.5

Which of these building is the highest?

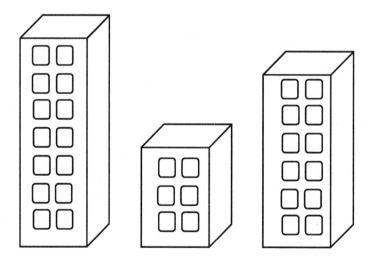

Figure 4.3-1: Building Height

Possible Solutions

Strategy: *Visualize Spatial Relationships*

In this strategy, a straight-edged object can be used to compare the heights of the buildings. If you use an index card or a piece of paper to cover the buildings from their bases, the last remaining building you see would be the highest. If you place a straight-edged ruler across the top of one of the buildings, you will be able to see where that building meets the others.

Strategy: *Analyze the Units*

The child may decide to count each window. If this strategy is proposed, ask the child if he can tell by looking. Follow-up questions: How many windows are there on the front of each building? How do these numbers compare?

Problem 4.6

A tessellation is defined as a plane surface covered in tiles that do not overlap and with no gaps. How many different regular polygons will successfully tessellate? What are some different quadrilaterals that will also tessellate?

A regular polygon is one whose side lengths are all equal in length (a square is a regular quadrilateral) and whose angles are all equal in measure.

Strategy: *Produce a Model*

In this strategy, multiple pieces of the same shape are necessary. You might use pattern blocks or create and cut out homemade shapes. Place the same shapes right up against each other, doing your best to ensure there is no space between the shapes at all. See if you can make a pattern by rotating or moving the shapes that will ensure that no space remains.

Strategy: Draw a Diagram

In this strategy, only one of each shape is necessary, as they can be traced on a piece of paper. Take a shape or create one by cutting out of an index card. Trace the shape on the paper one time. Trace the shape adjacent to the first shape, ensuring no spaces.

Possible Solutions

All squares tessellate.

All parallelograms tessellate.

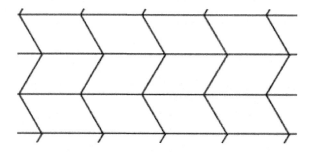

All triangles tessellate.

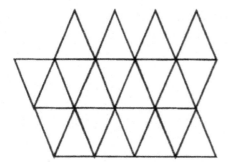

All quadrilaterals tessellate.

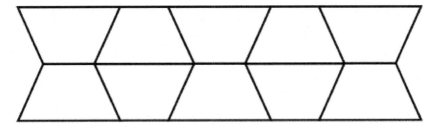

Regular pentagons do not tessellate by themselves.

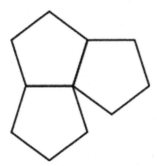

A follow-up question for this problem could be: What are Archimedean tessellations? What are some possible examples and non-examples of Archimedean tessellations?

Problem 4.7

Farmer Joe has 6 animals; some are cows and others are chickens. If he can count 18 legs, how many of each animal might he have?

Strategy: Eliminate Possibilities

In this strategy, take the maximum number of possible cows. Since there are 18 legs total, there can be no more than 4 cows. There are no "half-animals" allowed, so the number of chickens can be calculated based on Farmer Joe having 4, 3, 2, or 1 cows.

Four cows mean that 16 legs have been accounted for, so there would be only 1 chicken in this scenario. Three cows mean that twelve legs are accounted for, so there must be three chickens. Two cows mean that eight legs have been accounted for, so there must be 5 chickens. Finally, if there is only one cow, then there must be seven chickens.

Strategy: Produce a Model

In this strategy, we can take items to represent our chickens and cows. They can be objects that match the number of legs of each animal. For example, you might have toy animals or dinosaurs with four legs that can represent our cows. And you might use dolls or toy people or dinosaurs on two legs to represent the chickens. The combinations of models can be arranged until all of the possibilities have been discovered.

Possible Solutions

Since chickens have two legs and cows have four legs, the goal is to find out how many of each can add up to exactly 18 legs. Here are all of the possibilies:

Chickens	Chicken Legs	Cows	Cow Legs	Total Legs
1	2	4	16	18
3	6	3	12	18
5	10	2	8	18
7	14	1	4	18

Problem 4.8

One school limits the number of students in a class to no more than 30. A class can be grouped in equal groups of two, three, four or 12. How many students could be in the class?

Strategy: Produce a Model

For this problem, any type of object can be used as a model since all the items represented are the same. If you have 30 marbles or blocks or crayons, they can be used to represent the equal groupings. The objects can be arranged in groups of two, three, four and 12 to determine which numbers can be grouped with no extra objects.

Possible Solutions

This question relates to factoring skills. Since the class can be grouped into 12 students, the total number of students must be a multiple of 12. However, since the class is limited to 30 students, the only multiple of 12 that is also less than 30 is 24. This number of students can be evenly grouped into groups of two, three, and four students.

Problem 4.9

I have pennies, nickels, and dimes in my purse. If I have three coins, how much money do you think I have?

Possible Solutions

Strategy: Act It Out

You might provide the child with coins or allow them to take some coins out of the piggy bank in order to touch and work with the question from this problem. They can use the clues to determine which coins to pull out.

Strategy: Make a Systematic List

Three coins can be represented on paper by letters or numbers. The letter possibilities might look like this, where p = penny, n = nickel, and d =

dime: p n d = 1 + 5 + 10 = 16 cents. The child may represent each coin with a number and let p = 1, n = 5, and d = 10, writing 1 + 5 + 10.

An extension to this problem would be to ask: I have four coins. I have five coins. How much money could I have? What is the least amount of money I could have? What is the most amount of money I could have?

Possible Solutions for four coins

p p n d = 17 cents
p n n d = 21 cents
p n d d = 26 cents

Possible Solutions for five coins

p p p n d =18 cents
p p n n d = 22 cents
p p n d d = 27 cents
p n n n d = 26 cents
p n d d d = 36 cents

4.4 Everyday Problem Solving for Early Childhood (5, 6, & 7-Year-Olds)

Activity 4.1 Fact-finding

Encourage intellectual autonomy. When encountering problems, discussion and questioning should be the primary tools to direct learning as opposed to an adult administering knowledge. Interdisciplinary topics sometimes overlooked in the curriculum can be great explorations for adult/child pairs. Some of the following suggestions may even be concepts that the adult can learn even more about if researching with a child. Consider doing an internet search or library visit to explore the following:

- o Fibonacci
- o Electricity
- o Solar system
- o Pyramids
- o Moon and Tides

Activity 4.2 "Count on" Dice Game

Roll two dice. Start from one dice and use the counters on the second die to count forward. This can be practiced during traditional board game play or independently. For variation: This game can be played with die with more than six sides or more than two die.

For example, if the dice were rolled as follows, 3 and 6, you would read "3, 4, 5, 6, 7, 8, 9." Where 3 is read from the first die and then six of the following numbers are called.

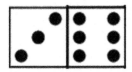

In another case, if the following were rolled (six then two), you would read "6, 7, 8."

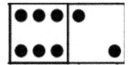

Activity 4.3 "Find the Missing Number"

This game can be played on a chalkboard, whiteboard, or paper. Take turns creating number sentences with a missing number. The number can be covered with a sticky note or paper, or could be represented as a box. Here are some examples

$$23 + \square = 55$$
$$\square - 10 = 30$$

Variation: Include multiple boxes. Challenge each other to figure out how many different ways you can fill in the boxes. Determine which number sentences have a writable number of answers and those we might go on writing answers for a very long (infinite) amount of time.

$$\square - 10 = \square$$
$$\square + \square = 14$$

Activity 4.4 Predict My Pattern

Use pattern blocks, bricks, coins, pipe cleaners, or any other set of objects to create patterns. Challenge each other to continue the pattern. Variations: Assign the patterns numbers. Create a pattern that can have more than one answer. Here is an example of a pattern that might have more than one answer: 0, 15, 30, ____, _____, _____.

One possible answer might be 0, 15, 30, 45, 60, 75,.... This pattern is developed by adding 15 to each number. Another solution might be 0, 15, 30, 45, 90, 105,... This pattern is achieved by adding 15 then multiplying by 2 (+ 15, x 2). One more option for this pattern might be 0, 15, 30, 45, 0, 15, 30, 45.... This pattern is developed using the clock and noting the location of the minute hand every 15 minutes.

Activity 4.5 Riddle Me Number

Provide clues to a number you are thinking of for your child to guess. Examples of clues you might give:
- *The number is odd and less than your age.*
- *There are ten desserts plated at Grandma's. There are two types of desserts. How many of each dessert?*
- *A pet store owner has twice as many birds as snakes. If he has 12 total birds and snakes, how many of each?*
- *There are an odd number of blocks. You can use all of the tiles to make three towers that are all the same size. How many blocks are there?*

Challenge: Your child determines the clues for you to guess!

Activity 4.6 How Many Ways?

Use blocks or interlocking bricks to create different ways to represent numbers. One study found this activity can reinforce understanding of place value and the idea that numbers can be represented in different formations. Here are some ways to represent the number eight using blocks:

You can challenge each other to create tens as one unit. Here are some examples of the number 24:

Activity 4.7 Involvement in Home Keeping

At this age you may find that your child is eager to help you. Consider delegating some age-appropriate tasks, especially when doing them together. Children at this age are capable of clearing dishes, watering plants, picking-up play areas, making their bed, preparing snacks or trays, sorting silverware, wiping shelves or cabinets, and setting the table.

Chapter 5

Grade School Gurus

5.1 Development of Grade School Children

The word "guru" comes from the Sanskrit word denoting a master or a teacher. Children at this age can now justify their reasoning. A key development in problem solving is being able to justify steps and results. While at age 3 and 4 children are seemingly filled with an infinite quantity of "Why" questions, at this stage you can turn the question — a very important one — back to the child. Asking children to defend their thinking will develop their ideas further and allow them to become comfortable explaining themselves, an important skill for problem solving and for a productive life.

As children enter 3rd grade, they begin to develop increasingly concrete thinking. Age 8 continues development in Piaget's concrete operational stage. In this stage children begin applying logic and reasoning to concrete events. Symbolic thinking is developed in this age range as well. While you are asking your child to substantiate her thinking, she might also be curious about concrete incongruities she encounters. Self-criticism also begins to be part of normal thinking.

Conservation is also more understood at this age. In other words, 8 year olds can comprehend that the identity of a described or understood measure or item remains the same if only superficial aspects are changed. An 8-year-old can understand that a stacked pile of 8 pennies, for example, has the same count as the same 8 pennies in a concurrent row or a row of pennies with a space or even the same pennies in a circle or pile. This is also true of the conservation of volume. Children can now understand that pouring liquid from a tall, slim glass into a wide, short glass doesn't change the amount of water. Decentration is another development in this stage. Children can take on another person's point of view and consider multiple dimensions. While children are increasing in cognitive demand, one caveat would be to refrain from heading into hypothetical thinking. Abstract thinking is not developed until later years and children may become confused if presented with scenarios or ideas at this level.

At this age, the importance of peer and social interaction increases. Children are aware of how they stand with their peers and how they are perceived becomes important. Social interactions change, in that what others think, particularly peers, begins to matter more and more. Children are also beginning to face peer pressure and the consequences of being part of a group or class. Troubleshooting friendships becomes necessary, as feelings are hurt and differences noticed. Children are now in need of help and direction about how to forgive and how to see another person's perspective, how to be both a good sport, whether in the position of a winner or a loser on a team. Children learn to encourage or discourage other with their words and actions.

5.2 Mathematics for Grade School Children

Students at this age are studying multiplication, division, and the relationship between the two. Students should fluent in the multiplication facts which result in products through 100. Students should be familiar with multiple ways to represent multiplication and division problems. Some examples of ways to represent these operations:

- area model
- counters
- tallies or marks
- arrays

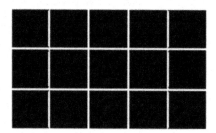

Figure 5.2-1: Area model representing 5 x 3 = 15

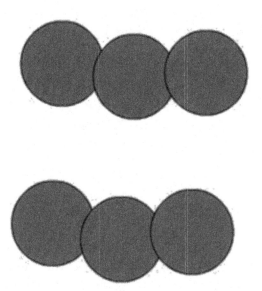

Figure 5.2-2: Counters representing 3 x 2 = 6

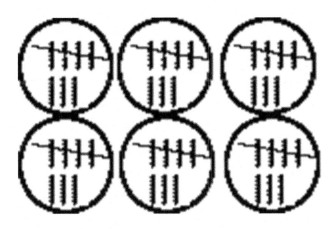

Figure 5.2-3: Tallies representing 8 x 6

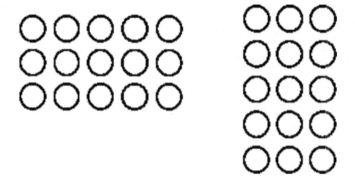

Figure 5.2-4: Arrays representing 3 x 5 and 5 x 3

Children of this age should also be identifying and explaining even more complex patterns. Fractions are also developed at this stage. Children are studying the unit fraction (1/a) and how the numerator and the denominator interact and what they represent. Fractions can be represented in several ways as well:

- on the number line
- as a portion of a shape: circle, square, triangle, etc.
- on a tape diagram
- using fraction bars

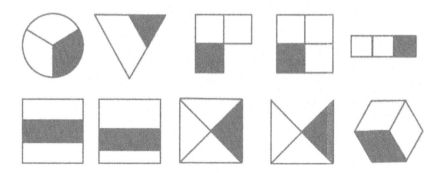

Figure 5.2-5: Fractions as portions of shapes

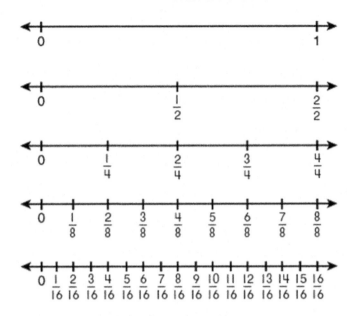

Figure 5.2-6: Fractions on the number line

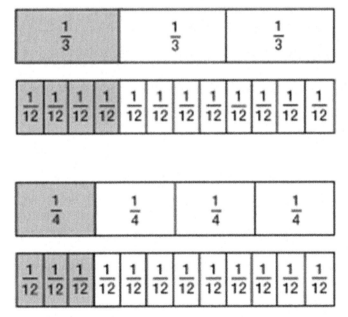

Figure 5.2-5: Equivalent Fractions using Tape Diagrams

1											

| $\frac{1}{2}$ | | | | | | $\frac{1}{2}$ | | | | | |

| $\frac{1}{3}$ | | | | $\frac{1}{3}$ | | | | $\frac{1}{3}$ | | | |

| $\frac{1}{4}$ | | | $\frac{1}{4}$ | | | $\frac{1}{4}$ | | | $\frac{1}{4}$ | | |

| $\frac{1}{5}$ | | $\frac{1}{5}$ | | $\frac{1}{5}$ | | $\frac{1}{5}$ | | $\frac{1}{5}$ | | | |

| $\frac{1}{6}$ | | $\frac{1}{6}$ | | $\frac{1}{6}$ | | $\frac{1}{6}$ | | $\frac{1}{6}$ | | $\frac{1}{6}$ | |

| $\frac{1}{8}$ | $\frac{1}{8}$ | $\frac{1}{8}$ | $\frac{1}{8}$ | $\frac{1}{8}$ | $\frac{1}{8}$ | $\frac{1}{8}$ | $\frac{1}{8}$ | | | | |

| $\frac{1}{9}$ | $\frac{1}{9}$ | $\frac{1}{9}$ | $\frac{1}{9}$ | $\frac{1}{9}$ | $\frac{1}{9}$ | $\frac{1}{9}$ | $\frac{1}{9}$ | $\frac{1}{9}$ | | | |

| $\frac{1}{10}$ | $\frac{1}{10}$ | $\frac{1}{10}$ | $\frac{1}{10}$ | $\frac{1}{10}$ | $\frac{1}{10}$ | $\frac{1}{10}$ | $\frac{1}{10}$ | $\frac{1}{10}$ | $\frac{1}{10}$ | | |

| $\frac{1}{12}$ | $\frac{1}{12}$ | $\frac{1}{12}$ | $\frac{1}{12}$ | $\frac{1}{12}$ | $\frac{1}{12}$ | $\frac{1}{12}$ | $\frac{1}{12}$ | $\frac{1}{12}$ | $\frac{1}{12}$ | $\frac{1}{12}$ | $\frac{1}{12}$ |

Figure 5.2-6: Fraction Bars

At this age, students should begin to recognize and become fluent in the use of rulers, using the whole number measures along with halves and fourths. They will begin to explore perimeter at this age, so they can use rulers to begin to measure around plane figures. They should be practicing drawing picture and bar graphs to represent data. They should also be able to interpret bar graphs, answering questions about differences in categories of data.

Table 5.2-1: Terminology for Grade School Children

Practice identifying and defining the following terms:		
>, <	Mass	Quadrilateral
Angle	Measurement	Quotient
Area	Multiple	Ray
Associative Property of Multiplication	Multiply	Rhombus
Commutative Property of Multiplication	Order of operations	Round
Coordinate plane	Parallel	Ruler
Decimal	Parenthesis/Brackets	Square units
Distributive Property	Perimeter	Symmetry
Divide	Perpendicular	Time to the nearest minute
Equal parts	Picture graph	Unit fraction
Equivalent fractions	Place value	Unknown
Factor	Polygon	Volume
Line segment	Product	

5.3 Sample Problems for Grade School Children

Problem 5.1

You would like to enclose a garden with some leftover fence. The length of the fence is 18 feet, and the fence is cut in one foot lengths. What are some of the ways you can make the garden if you would like the garden to be in the shape of a triangle and only have whole number dimensions?

Strategy: Produce a Model

For this strategy, it would be useful to use straight edged cards, rulers, rods, or straws that can be made or bent into lengths of the numbers 1 through 16. The child can choose 3 addends that sum to 18 and see if the lengths make a triangle. Then try choosing 4 lengths to see if a quadrilateral is possible. Record both the successes and the unsuccessful attempts to avoid repeating.

Possible Solutions

The important thing to remember about triangles is that the triangle inequality dictates which triangles are possible. Namely, the sum of the lengths of two sides of a triangle must always be longer than the third side. Otherwise the three vertices will not meet. This will be evident, for example, when trying to use side lengths of 4 and 5 and 9. The lengths sum to the required 18, however 4 and 5 will never meet the length of 9 at any angle. Notice that $4 + 5 = 9$. So the two sides will be stretched to make an 180° angle and then no longer is it a triangle. If the lengths of 2, 6, and 10 are attempted, they again meet the 18 length requirement however only one leg will reach the length of 10. One length will always be too short.

The possible triangles are:

8	8	2
7	7	4
5	7	6
6	6	6
4	7	7
5	6	7
2	8	8
3	7	8
4	6	8
5	5	8

Problem 5.2 The concession stand at a baseball game has the following menu. If you have $3.00, how many different ways can you spend exactly that amount?

Item	Cost
Popsicle	$.50
Gumball	$.25
Water	$1.00
Candy bar	$1.50

Strategy: Act It Out

You might use real or play money and have someone act as the customer and someone else act as the cashier at the concession stand. If you have or can make objects that can represent each item for sale, the process of money changing hands in exchange for items can make the experience real.

Strategy: Make a Systematic List

Start with the least or most costly item. Determine how many different purchases can be made with it. Move to the next most expensive item, create a new column including only it and less expensive items, and so on until you have exhausted all items and possibilities.

Possible Solutions

Let P = popsicle,   G = gumball,   W = water,   C = candy bar

2C	3W	6P	12G
C + W + P	2W + 2P	2P + 8G	
C + W + 2G	2W + 4G	3P + 6G	
C + 6G	2W + 1 P + 2G	4P + 4 G	
C + 3P	1W + 4P	5P + 2G	
C + 2 P + 2 G	1W + 8G		
C+ P + 4G	1W + 3P + 2G		
	1W + 2P + 4 G		

Problem 5.3

Two sisters watched a group of cyclers ride by. The younger sister counted 7 riders. The older sister counted an odd number of wheels. What is the range of the number of tricycles that could have been in the group?

Possible Solutions

Strategy: Make a Systematic List

In this strategy, we will use the given information to analyze all of the possible types of bikes. There were 7 riders, so we know there were 7 cycles. If there were an odd number of wheels, there was at least 1 tricycle.

Number of Bicycles	Wheels per Bicycle	Number of Tricycles	Wheels per Tricycle	Total number of wheels
0	2	7	3	21
1	2	6	3	20
2	2	5	3	19
3	2	4	3	18
4	2	3	3	17
5	2	2	3	16
6	2	1	3	15

Given this information, we see that there could have either been 7, 5, 3, or 1 tricycle in the group because the total number of wheels will result in an odd number. It is possible there were seven tricycles but there had to be at minimum of one to ensure an odd number of wheels.

Problem 5.4

Use the associative property to demonstrate all of the ways you can sum three numbers to equal 15.

Possible Solutions

Strategy: Make a Systematic List

There are a number of ways that we can create the sum of 15 using three numbers. If we start with the two smallest digits, we can develop the following list:

1 + 1 + 13
1 + 2 + 12
1 + 3 + 11
1 + 4 + 10
1 + 5 + 9
1 + 6 + 8
1+ 7 + 7

You may notice that after this the numbers would begin to repeat (1 + 6 + 8 are the same addends as 1 + 8 + 6). So, let's move on to addends that include 2:

2 + 2 + 11
2 + 3 + 10
2 + 4 + 9
2 + 5 + 8
2 + 6 + 7

Again, we have reached a point where the numbers would repeat. So, we move on to 3:

3+ 3 + 9
3 + 4 + 8
3+ 5 + 7
3+ 6 + 6

And our set of addends with 4 or higher:

4 + 4 +7, 4 + 5 + 6, 5 + 5 + 5

Now, to demonstrate the associative property, apply parenthesis to the above addend lists:

(1 + 1) + 13	1+ (7 + 7)	(3 + 5) + 7
1 + (1 + 13)	(2 + 2) + 11	(3+ 6) + 6
(1 + 2) + 12	(2 + 3) + 10	3+ (3 + 9)
1 + (2 + 12)	(2 + 4) + 9	3 + (4 + 8)
(1 + 3) + 11	(2 + 5) + 8	3+ (5 + 7)
(1 + 4) + 10	(2 + 6) + 7	3+ (6 + 6)
(1 + 5) + 9	2 +(2 + 11)	(4 + 4) +7
(1 + 6) + 8	2 + (3 + 10)	(4 + 5) + 6
(1+ 7) + 7	2 +(4 + 9)	(5 + 5) + 5
1 + (3 + 11)	2 + (5 + 8)	4+ (4 + 9)
1 + (4 + 10)	2 + (6 + 7)	4+ 4 +7,4 + (5 + 6)
1 + (5 + 9)	(3+ 3) + 9	5 + (5 + 5)
1 + (6 + 8)	(3 + 4) + 8	

Wow! We've made 38 options that all sum to 15.

Problem 5.5

Use the following diagram to create fractions and explain how they are represented in the diagram.

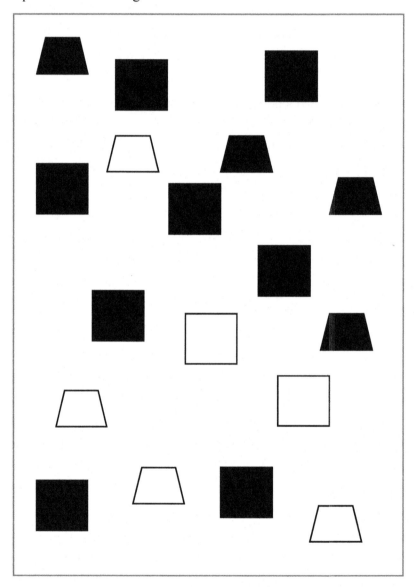

Strategy: Organize information

There are a number of fractions that we can make from the diagram. We can begin with the squares:

$\frac{2}{10}$ or $\frac{1}{5}$ there are two white squares out of ten total squares.

$\frac{8}{10}$ or $\frac{4}{5}$ there are eight black squares out of ten total squares.

$\frac{4}{8}$ or $\frac{1}{2}$ there are four white trapezoids out of eight total trapezoids.

$\frac{4}{8}$ or $\frac{1}{2}$ there are four black trapezoids out of eight total trapezoids.

$\frac{6}{18}$ or $\frac{1}{3}$ there are six white shapes out of eighteen total shapes.

$\frac{12}{18}$ or $\frac{2}{3}$ there are twelve black shapes out of eighteen total shapes.

5.4 Everyday Problem Solving with Grade School Children

Since peer interaction and friendships are important, engage with the child and their peers within social structures and situations. Host game nights with kids or other families. Engage in games on a regular basis at home. The games can be store-bought such as board or card games, or they can be traditional games like "Charades" or ones that specifically develop on-grade-level numeracy. Here are some examples:

Activity 5.1 "What Whole Number Comes Next?" Game

Develop number sense while playing this game while you're on a walk or taking a trip in a car. If someone spies a number, s/he can call out that number. To begin, this supports the development of fluency of how to read multi-digit numbers. Another player takes a turn by calling out the next larger whole number. You might find numbers on houses or road signs, on license plates, on clothing or in store displays. As an adult, you might be sure to pick two or three digit numbers. Take note of consistent mistakes for supporting your child at home and work with them to troubleshoot if the number is unfamiliar. For example, if you see "860" as an area code and call it out, you might ask your child to solve a smaller problem within this number: "What comes after 60?"

> *"Countdown" Variation*: This game has the same idea, however seek out numbers up to 20, and the group takes turns counting down from that number.
> *"Rounding" Variation*: Children at this age are learning to round to the nearest tens or hundreds place. After calling out a number, round to the nearest ten or hundred and explain why.
> *"More or Less" Variation:* Find two numbers to compare. You might see two numbers on sports jerseys or costs on a menu: Is 23 more or less than 32?

Activity 5.2 "Ordinal or Cardinal" Game

This can be a spontaneous game, where instances of number come up while at a sports event, while reading or watching a show. You can ask a child to determine if the number is ordinal or cardinal and explain why. Ordinal numbers explain position (such as first, second third) while cardinal numbers indicate the quantity in a set (0, 1, 2, 3...)

The child can also ask you the same: "ordinal or cardinal"? You can make it fun by even saying an incorrect answer and allowing your child to correct or teach you or other members of the family or group.

Here are some instances where you might quiz each other:

- At a sibling's soccer game: "Goal! The score is now 5 to 0."
- While cheering on a marathon: "Jimmy now takes the lead and is in 23rd place."
- Watching election results or Olympic games.

Activity 5.3 "I'm thinking of a number" Game

Providing clues to participants of this game helps to develop numeracy. Children at this age are learning terms to qualify or describe numbers. Some of these terms you might use: fewer, greater, about, near, between, etc.

Activity 5.4 "I Spy" Geometry Game

Create "Bingo" cards with the names of geometric shapes. As you head on a trip or go out for a walk, seek out instances of plane closed figures along the way. Examples of figures your child should be able to identify and describe attributes: square, rectangle, rhombus, quadrilateral, circle, triangle, polygon, kite, octagon, hexagon, pentagon, parallelogram, trapezoid.

Activity 5.5 Make & Evaluate Predictions

While watching a movie, TV show or reading a book or story, stop and discuss possible consequences. Make predictions together and, after, develop problem solving or scenario games.

Activity 5.6

Create cards with age-appropriate issues your child will face at this time, you can pull them out of a hat and discuss or act out possible reactions or solutions. Some issues your child might face at this time: students are teasing or bullying a child at school, a friend asks to copy homework, lying about a mistake or broken rule, taking someone's belongings, being asked to be a girl/boyfriend, tattling, whining, exaggerating, how to react when others are rude or disrespectful, etc.

Activity 5.7

Talk about growth and change with your child. Set obtainable goals. Discuss what actions will lead to success in the goals. Track and monitor progress toward their goals.

Activity 5.8

Support and develop your child's interests. Take them to an array of sporting events, museums, zoos, parks, workplaces, fairs, festivals, etc. Provide opportunities to build and create based on interest.

Activity 5.9

Make an explicit effort to understand their point of view. Model and discuss acceptance of differences of opinion.

Activity 5.10

Distinguish between fact and opinion. Discuss how to treat those who are different and who think different. Put yourself, and teach your child to put him/herself into the perspective of someone else.

Activity 5.11

Allow them to engage in a process of negotiation with you, both in real situations and mock situations

Activity 5.12

Volunteer. Purposely engage with folks from different cultures, socio-economic classes, faith, political beliefs, etc.

Activity 5.13

Be sure to have children take on additional responsibilities at home as they grow. They are capable of more at each age, and should be encouraged to be part and contribute to the family unit. Consider distributing some of the following chores at this age: fold towels, rake leaves, make salad, replace towels / napkins / toilet tissue, gather trash, fold laundry, match socks, peel vegetables, load dishwasher, sweep, wipe tables or dust.

Activity 5.14

Draw each of the following using a different shape or model or object: halves, thirds, fourths, sixths, and eighths. Create fraction flash cards or a game where people have to draw fractions such as two thirds of five twelfths.

Activity 5.15

Create activity cards using the terminology listed in this chapter. Put the cards into an envelope or card box. Choose a term at random. Create or find an activity to reinforce the term. For example, if the term is a geometric figure, find as many of that figure as you can. Alternatively, search the term in a browser or app to find programs to learn more about it.

Chapter 6

Middle School Minds

6.1 Development in the Middle Grades

Exiting the concrete operational stage, children at this age are beginning to enter the formal operation period. They are able to conserve number, mass, volume as well as think logically. They are able to choose and apply methods, use deductive logic and strategic thinking. Abstract thinking and making predictions are also a hallmark of this age. The scientific method is more readily applies, as well as categorization and grouping of items by concurrent features (such as listing rookie players on different teams by their graduation year or separating utensils by both plastic versus silver and by type). The consequence of this thinking, on the other hand, is the onset of the questioning of authority and societal norms.

At this age, the present time is of utmost importance. There is a lack of ability to see long-term consequences and to think about the future in any real way that would impact thinking or action. Everything is very concrete: either good or bad, amazing or terrible.

Since egocentrism is beginning to dominate the child's mind at this age, it is important to balance out the feelings of self-importance with experiences with empathy and sympathy. There are increased risk-taking impulses, however impulse control is not quite developed for another 10-15 years, so it is also important to provide safe space for children to take risks and evaluate consequences. Metacognition is also being developed at this time, and can be key to supporting the problem-solving, academic and emotional development of the child.

6.2 Mathematics in the Middle Grades

At this time, students are beginning to apply ratio reasoning to solve problems. They will be building upon their understanding of multiplication and division to apply these operations to fractions. Finding common factors and multiples will be required of them, and will be helpful in solving problems including fraction operations. Numeracy is developed through the application of positive and negative numbers as well as absolute value. Students will be applying absolute value to find distance in the coordinate plane.

Algebraic thinking becomes more fully developed as children are writing, evaluating, and finding equivalency with expressions that include variables, coefficients (both positive and negative) and exponents. Precision in mathematical terminology is expected as well when identifying operations and components of mathematical problems. They will also be grappling with single-variable equations and inequalities. Finally, they will be engaging with relationships between dependent and independent variable quantities.

At this time, geometry will involve real-world problems applying the formulas for area, surface area, and volume (oftentimes applying algebraic thinking). And finally, students will be required to master understanding of statistical variability, including the ability to summarize and describe distributions.

Table 6-1: Terminology for Children in Middle Grades

Practice identifying and defining the following terms:		
Absolute value	Independent variable	Ratio
Adjacent angles	Inequality	Rational number
Bivariate data	Inference	Reflection
Categorical data	Interquartile range	Rotation
Circumference	Irrational number	Sample population
Coefficient	Mean	Sample space
Congruence	Mean absolute deviation	Scatter plot
Construction	Median	Scientific notation
Dependent variable	Mode	Simple interest
Dilation	Negative numbers	Simulation
Divide fractions	Net (of three-dimensional figures)	Slope
Dot plot	Outcome	Supplementary angles
Equation	Percent	Surface Area

Equivalent Ratio	Percent increase and decrease	Term
Evaluate	Probability	Translation
Exponent	Probability model	transversal
Expression	Proportion	Unit Rate
Exterior angle	Pythagorean Theorem	Variability
Frequency	Quadrant	Vertical angles
Function	Quartile	Vertices
Greatest common factor	Radical	y-intercept

6.3 Sample Problems for the Middle Grades

Problem 6.1

You've been asked to design a custom cover for tablet for a high school whose school colors are purple, green, and white. The cover will have horizontal stripes of the same width. There must be 2 purple stripes, 2 green stripes, and one white stripe. If you are going to show your client all possible designs?

Possible Solutions

Strategy: Produce a Model

Your visual learner might head over to the construction paper pile, cut 5 strips, and begin to lay them out accordingly.

Strategy: Act it Out

Your kinesthetic learner might ask 5 family members to each hold a letter or paper to represent each color. The child can re-arrange the family members to represent each. The child might associate the colors with people: if brother and aunt are both holding green, they can never stand next to each other.

Strategy: Eliminate Possibilities

One way to determine the solution to this is through "opposites." A child might figure out how many flags will **not** work, and subtract this number from how many flags *are* possible.

Strategy: Make a systematic list

Abbreviating "P" for purple, "W" for white and "G" for green can help in the development of a systematic list, where the possibilities are listed:

PGWGP
PGWPG

Strategy: Draw a Diagram

In this case, a tree diagram can be of help to ensure that the rules are applied. One color can be started and filled out until all of the possibilities are exhausted:

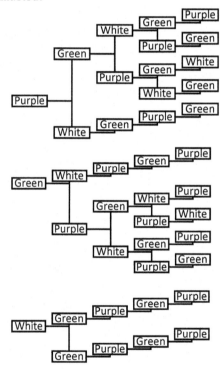

Problem 6.2

Find three fractions between five eighths and three fourths. Create visual representations to demonstrate that they are between the two given fractions.

Possible Solutions

Strategy: Produce a Model

Take a piece of paper and fold it into eighths. Or take eight or 16 of the same item such as interlocking toy bricks to demonstrate the fraction. Then, use the figures or items to break down the fraction into smaller parts (such as folding the paper again or adding additional objects).

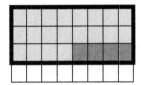

5/8 or 10/16 or 20/32 (light gray)
Fractions in between: 21/32, 22/32, 23/32
(dark gray + light gray)
¾ or 6/8 or 12/16 or 24/32 (boxed)

Strategy: Visualize Spatial Relationships

In this problem, a number line might be helpful. To see the breakdown of a number into smaller parts and how the fractions relate to each other might be very helpful.

$$\frac{1}{16} \quad \frac{1}{8} \quad \frac{3}{16} \quad \frac{1}{4} \quad \frac{5}{16} \quad \frac{3}{8} \quad \frac{7}{16} \quad \frac{1}{2} \quad \frac{9}{16} \quad \frac{5}{8} \quad \frac{11}{16} \quad \frac{3}{4} \quad \frac{13}{16} \quad \frac{7}{8} \quad \frac{15}{16} \quad 1$$

Here we can see that $\frac{11}{16}$ falls between $\frac{5}{8}$ and $\frac{3}{4}$ on the number line. There are also two more notches that can be determined using half of the sixteenths, or thirty-seconds. If we convert $\frac{5}{8}$ to its equivalent of $\frac{20}{32}$, the next fraction will be $\frac{21}{32}$ on which is another fraction found within our range. Using that same procedure, the fraction after $\frac{11}{16}$ ($\frac{22}{32}$) would be $\frac{23}{32}$.

Problem 6.3

One day in a college on three campuses whose enrollments are in the ratio of 1:2:5 and exactly one in 25 students were absent in every school. If there were no less than 15,000 students in the district. How many students were present in each school?

Strategy: Use Deductive Logic

Using the clues from this problem, we know that the number of students on each campus will be divisible by 1, 2, 5, 8 (which is 1+2+5) and 25. If we broke the 15,000 students into 8 groups and then distributed 1 group to the first campus, 2 of these groups to the second campus, and the remaining 5 groups to the third campus, their ratios would demonstrate 1:2:5. After this distribution, dividing each result by 25 will calculate the numbers of absent students.

Strategy: Organize Information

In this strategy, create a table to help determine the number of students on each campus. This first diagram represents the portion of students on each campus based on the given ratios:

Campus 1	Campus 2		Campus 3			
15,000 Students Total						

This chart can help organize the number of absent students on each campus:

	Campus 1	Campus 2	Campus 3	College
	Ratio: 1	Ratio: 2	Ratio: 5	Ratio: 8
Enrollment				
Absentees				
Present Students				

Possible Solutions

	Campus 1	Campus 2	Campus 3	College
	Ratio: 1	Ratio: 2	Ratio: 5	Ratio: 8
Enrollment	1875	3750	9375	15000
Absentees	75	125	375	600
Present Students	1800	3625	9000	11250

Problem 6.4

An owner is making an enclosed pen out of fence. If the dimensions must be whole numbers, what is the largest pen that can be made out of 24 feet of fence?

Possible Solutions

Strategy: Create a Physical Representation

Choose toothpicks or crayons or another straight item of equal length that you can use to model the situation. Use the crayons to demonstrate the possible areas from the given length.

This represents the longest possible area using 24 lengths. The width is one and the length is 11 which makes an area of 11.

This is another option. The width is two and the length is 10 which makes an area of 20.

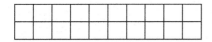

This area has a width of 3 and a length of 9 and the area totals 27.

Another possibility is a width of 4 and length of 8 with an area of 32.

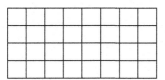

The final option which has the largest area is the six by six with an area of 36.

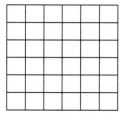

Strategy: Make a Systematic List

The smallest width would be 1 since the dimensions must be whole numbers. If we have a width of 1, then 2 feet of fence are taken up already and 22 feet remain for the length. Since a rectangle has two of the same side lengths, they would both be 11 feet. We can

Width	Length	Area
1	11	11
2	10	20
3	9	27
4	8	32
6	6	36

Problem 6.5

There is a bargain book sale with books in boxes by genre. There are six mystery books, five science fiction books, and three nonfiction books. How many different ways can you choose one of each genre?

Strategy: Identify Subproblems

In this strategy, we can explore what would happen if there were fewer books or fewer kinds of books. What if there were only two types of books? What if there were only one of each? Two of each? How do these results relate to each other?

Strategy: Make a Systematic List

In this strategy, we can label the books and create a list of all of the possibilities. For brevity, we can name the mystery books M1, M2, M3, M4, M5, & M6. And the science fiction books S1, S2, S3, S4, & S5. Finally, we can name the nonfictions books N1, N2, & N3. Choose one from each type and make a list of all of the options without repeating. Here is the list of possibilities if the first mystery book is chosen:

M1, S1, N1
M1, S1, N2
M1, S1, N3
M1, S2, N1
M1, S2, N2
M1, S2, N3
M1, S3, N1
M1, S3, N2
M1, S3, N3
M1, S4, N1
M1, S4, N2
M1, S4, N3
M1, S5, N1
M1, S5, N2
M1, S5, N3

Possible Solutions

The total number of options is equal to the product of all of the options. In this case, the options are equal to 6 x 5 x 3 = 90. There are 90 ways to choose one book from each category.

Problem 6.6

How many sixths are in two thirds?

Possible Solutions

Strategy: Produce a Model

In this strategy, we are **creating** tape diagrams to model the relationship between sixths and thirds.

One Third	One Third	One Third

One Sixth	One Sixth	One Sixth	One Sixth	One Sixth	One Sixth

Strategy: Draw a Diagram

Here is a circle cut into sixths, with two thirds shaded.

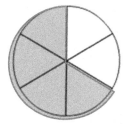

Problem 6.7

At a children's party, the magician asked for the product of the ages of three sisters. The dad said, "72." The magician then asked, "What is the favorite color of the oldest?" How would the answer to this question help the magician guess the girls' ages? How many possibilities are there for the girls ages? What other questions would help the magician get close to the correct answer?

Strategy: Eliminate Possibilities

In this strategy, we explore the options that could multiply to 72. This is a case of factorization and knowing the factors of 72.

Here are all of the factors of 72:

$$1, 2, 3, 4, 6, 8, 9, 12, 18, 24, 36, 82$$

The magician's question about the oldest helped him determine if there are any twins in the family. The magician might also ask if any of the girls' ages is prime or the favorite color of the youngest family member. The magician could also ask the sum of their ages to get closer to the answer.

Possible Solutions

The girls ages could be (although some do not seem reasonable)

1, 2, 36
1, 3, 24
1, 4, 18
1, 8, 9
2, 2, 18
2, 3, 12
2, 4, 9
2, 6, 6
3, 4, 6

Problem 6.8

Triangular numbers are integers that represent the number of items that can be represented in the form of an equilateral triangle. They can be represented by creating a triangular grid of points where the first row contains a single element and each subsequent row contains one more element than the previous row.

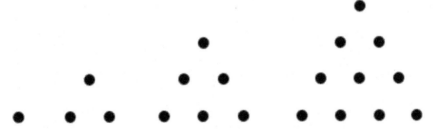

Figure 6.3-1: Triangular Numbers

List the next 5 triangular numbers.

Strategy: Draw a Diagram

You could use pen or pencil to depict the next five numbers. You also might just add the bottom row to a previous triangle in order to find the total for each triangle without starting from scratch.

Possible Solutions

The first nine triangular numbers are:
0, 1, 3, 6, 10, 15, 21, 28, 36, 45

Problem 6.8

A piggy bank includes at least one of each of the following coins: penny, nickel, dime, and quarter. There is the same number of pennies as quarters. There are twice as many dimes as quarters. There are three times as many nickels as dimes. What could be the makeup of the change in the piggy bank?

Possible Solutions

Strategy: Guess and Check

In this case, we will take a guess how many pennies are in the jar. Once we have entered the number of pennies, we can calculate the numbers of each coin and then calculate the total amount of money in the piggy bank. The only limit to the possibilities in this case is the size of the piggy bank!

Coin	Penny	Nickel	Dime	Quarter	Total Money
Value	$ 0.01	$ 0.05	$ 0.10	$ 0.25	
Ratio	= Q	=3D	=2Q	=P	
	100	600	200	100	$ 76.00
	35	210	70	35	$ 26.60
	79	474	158	79	$ 60.04
	15	90	30	15	$ 11.40
	250	1500	500	250	$ 190.00
	3000	18000	6000	3000	$ 2,280.00

Strategy: Make a systematic list

Since each of the other amounts of coins can be represented once the quantity of pennies and quarters are known, there are a number of solutions that can be determined by using calculations and the knowledge of numbers.

Coin	Penny	Nickel	Dime	Quarter	Total Money
Value	$ 0.01	$ 0.05	$ 0.10	$ 0.25	
Ratio	= Q	=3D	=2Q	=P	
	1	6	2	1	$ 0.76
	2	12	4	2	$ 1.52
	3	18	6	3	$ 2.28
	4	24	8	4	$ 3.04
	5	30	10	5	$ 3.80
	6	36	12	6	$ 4.56

6.4 Everyday Problem Solving with the Middle Grades

Activity 6.1

Ask your child to put mathematical terminology into his or her own words. Check for understanding of the terms s/he is expected to know. Seek ways to apply the words to your activities or routines. Some words to discuss at this age: area, volume, mass, surface area, term, coefficient, operation, exponent.

Activity 6.2

Analyze quantitative data you come across. Whether in an article, in a video or book, review the variability and the dependent and independent variables in the data set. Discuss the factors that may have influenced the data. Talk about why the data was important to the situation.

Activity 6.3

Take an in-depth look at the calendar. Have your child explore percentages and ratios by comparing parts of time. Here are some examples to try: What ratio or percentage of a year is weekend? What ratio or percentage of a year is a week? If a person spends 8 hours a day sleeping, what is the total number of sleeping versus waking hours in a week? Month? year?

Activity 6.4

Play factorization games. Challenge each other to come up with numbers with the highest number of factors. Determine all of the factors of numbers you encounter on trips or outings. Practice prime factorization as well.

Activity 6.5

Have your child plan an outing, vacation, or shopping trip using a certain budget. The child could research discounts and rates in order to ensure the most value for each dollar.

Activity 6.6

Create an unfair game. Explain which player you would like to be in the game and why.

Activity 6.7

At this time, children can take on even more responsibility in the home. Children can help by gardening, trimming bushes, or mowing the lawn, mopping or vacuuming floors, cleaning shelves and countertops, ironing clothes, painting and sewing buttons or socks. By the end of middle school, your child can clean a vehicle, bake cookies or bread, cook dinner, wash windows, build shelves from a kit, and bring in the mail.

Activity 6.8

Choose a news article that includes data or a study and analyze it together. Discuss the conclusions of the author of the study and the interpretations of the author of the article. Discuss who is benefitting from the study. Consider who funded the study. Talk about the potential flaws in the data.

Chapter 7

Teenage Thinking

7.1 Development of Teenagers (Ages 14-15)

There is no shortage of literature and oral history concerning the development of the teenager in the Western world. The changes during this period are significant and difficult for both parents and children as the new waters become difficult to navigate. They are seeking their own identity. Young people still crave the security and familiarity of their families but desire independence and freedom, creating a contentious conflict. In other civilizations, the transition from child to adult is swift and this period might include the child leaving the family home. It is during this age that humans develop the desire for relationships and begin to fall in love, replacing (or conflicting) with the norms set by the comfort of family.

Cognitive abilities increase in their complexity as the teenager begins to think philosophically and into the future. Analytic and argument skills are more developed, leading to new and more depth in questioning. Ethical and moral thinking is also emerging as important.

7.2 Mathematics for Teenagers (Ages 14-15)

During this time, students are typically enrolled in either an integrated mathematics course or separated Algebra and Geometry courses. In either case, the material within this grade band is very similar.

Teenagers are now being re-introduced to the Real Number System, and getting a better sense of the "big picture" of how the properties of numbers and the rules of numbers interact to create a system. They are comparing and categorizing types of numbers such as rational, irrational, whole, integer, and counting. They are applying reasoning to understanding how number types interact with each other. Algebraic thinking is being taught as a system that is analogous to the integers they have already become familiar with.

This curriculum introduces functions and the use of functions to model situations. Solving and graphing functions (and systems of functions) is necessary and is also foundational for further study. Fundamentals of probability and statistics are being taught to teens at this age as well. Understanding how data are represented and the likelihood of events are calculated are important foundational understandings.

In the realm of geometric thinking, the major areas of study at this age include recognizing congruence and similarity (and performing calculations based on theories). Properties of geometric figures including circles are applied to solving for dimensions.

7.3 Sample Problems for Teenagers (Ages 14-15)

Problem 7.1

A store is having an end of season sale. All merchandise from the previous season will be marked down 25%. Each Sunday, the merchandise will be marked down an additional 25%. Your friend is thinking that you should go shopping during the fourth week when items are free (four 25% discounts would mean 100% discount!) Do you agree? Explain why or why not.

Possible Solutions

Strategy: Produce a Model

If this bar represents the full price of the item:

After the first discount, the item will cost only three of the four bars:

| | | | 25% off |

The next week, 25% will be taken off the above, with only 75% remaining:

		25% off

In week 3, 25% will be reduced from the remaining price

	25%

In week 4, the price will be reduced 25% of the remaining price:

	25%

Strategy: Identify Sub-problems

In this strategy, we will take an example of an item with an original price of $100 and calculate the discount and discounted price each week.

	Current Price	25% Discount	Discounted Price
Week 1	$100	$25	$75
Week 2	$75	$18.75	$56.25
Week 3	$56.25	$14.06	$42.19
Week 4	$42.19	$10.55	$31.64

Strategy: Convert to Algebra

If you begin with a total cost, x, and you find the first discounted price, you can calculate:

$$x - .25x$$

to be the price after one week, which equals $.75x$. After one week, calculate 25% off the new price of $.75x$:

$$.75x - .25(.75x)$$
$$= .75x - .1875x$$
$$= .5625x$$

After two weeks, the price will be a little more than 56% of the original price. To calculate the price after three weeks, reduce the price again by 25% of the current price:

$$.5625x - .25(.5625x)$$
$$= .5625x - .140625x$$
$$= .421875x$$

After three weeks, the price will be a little over 42% of the original price. We repeat one more time to determine if the price will become 0% of the original.

$$.421875x - .25(.421875x)$$
$$= .421875x - .10546875x$$
$$= .31640625x$$

After four weeks of discounts, the cost will be about 31.6% of the original price.

Problem 7.2

In a group of 320 sports fans, 85 play basketball and 200 follow basketball. Sixty-five sports fans both play basketball and follow basketball. How many fans neither play or follow basketball?

Possible Solution

Strategy: Draw Venn Diagram

320 Fans

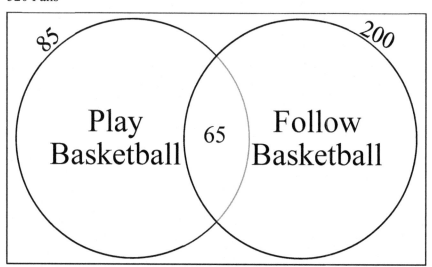

Use the Venn Diagram to calculate the number of students who play or follow basketball. There are 65 students that belong to both groups, so if we add 200 + 85, we must subtract the overlap of 65. This means 220 students will be accounted for within the circles who either play or follow basketball. Subtracting 220 from 320, we find that 100 fans neither play nor follow basketball.

Problem 7.3

The ratio of the lengths of two squares are 3:5. What are the ratios of their areas?

Possible Solution

Strategy: Look for a Pattern

In this strategy, we will compare the results of comparing side lengths to areas of squares within the ratio.

Length 1	Area 1	Length 2	Area 2
3	9	5	25
6	36	10	100
9	81	15	225
12	144	20	400
15	225	25	625

There is a discernable pattern. The ratio of the sides are squared to create the ratio of the areas. So if there is a ratio of the sides of a square that is a:b, then the ratio of their areas will be $a^2:b^2$.

Extension: Could this be true for triangles? Rectangles?

Problem 7.4

A farmer is burying a cylindrical fuel tank for propane gas. The dimensions read 72" x 48" x 48". If fuel costs $2.79 per gallon, how much will the tank cost to fill?

Possible Solution

Strategy: Draw a Diagram

The dimensions of the cylinder read 72"x48"x48". This would seem to indicate that two of the ends (which are identical) have a width of 48." Since the base of a cylinder is the shape of a circle, the diameter of the circle must be 48." The volume formula for a cylinder is $A=\pi r^2 h$. In this formula r is the radius of the circle and h is the height. The radius can be calculated by dividing the diameter in half. So, we can evaluate the expression with our given values:

$$A = \pi \left(\frac{48}{2}\right)^2 \cdot 72$$
$$A \approx 130{,}222.08 \text{ in}^3$$

There are 231 cubic inches in a gallon, so we must convert this to the units that will allow us to find the cost. If we divide: $\frac{130{,}222.08 \text{ in}3}{231}$, we will find approximately 563.7 gallons in the cylinder. At a cost of $2.79/gal, the total cost would be about $1,562.81.

Problem 7.5

Shirts and pants were on a "one price" sale. Judy purchased 7 shirts and 4 pants for $75. Kim purchased 12 shirts and 14 pants for $200. How much did one shirt cost? What was the price of one pant?

Possible Solutions

Strategy: Convert to Algebra

There are a number of ways to solve this system of equations, including graphing and substitution. If you use the substitution method, you might take *7s + 4p = 75* and *12s + 14p = 200* and solve one for *p*, then substitute into the other equation. For example:

$$7s + 4p = 75$$
$$4p = 75 - 7s$$
$$p = \frac{75 - 7s}{4}$$

Now substitute into the other equation:

$$12s + 14p = 200$$
$$12s + 14\left(\frac{75 - 7s}{4}\right) = 200$$
$$12s + \frac{1050 - 98s}{4} = 200$$

$$12s + 262.5 - 24.5s = 200$$
$$-12.5s + 225 = 200$$
$$-12.5s = -62.5$$
$$s = 5$$

Now we know if s = 5, then p must be 10 to make all of the equations work. The shirts were on sale for $5 and the pants were $10

Strategy: Produce a Model

Plot the following equations:

$$7s + 4p = 75$$
$$12s + 14p = 200$$

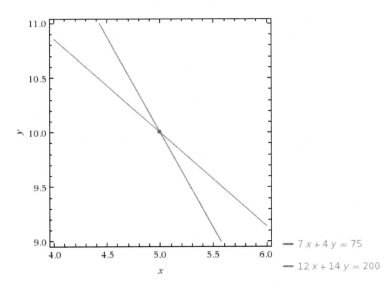

This graph indicates a point where the two equations meet, or the solution of the set. The solution lies at (5,10), indicating the costs of the items.

Strategy: Use Matrix Logic

To solve this using matrices, we will need to ensure that our equations are in standard form $(Ax+By=C)$. Let s = the price of the shirt and p = the price of the pants. The equations for this scenario are as follows:

$$7s + 4p = 75$$
$$12s + 14p = 200$$

Convert these to matrix form:

$$\begin{bmatrix} 7 & 4 & 75 \\ 12 & 14 & 200 \end{bmatrix}$$

Next, use matrix operations to change the first two columns to the identity matrix:

$$\begin{bmatrix} 1 & 0 & 5 \\ 0 & 1 & 10 \end{bmatrix}$$

7.4 Everyday Problem Solving with Teenagers (Ages 14-15)

Activity 7.1 Classic Circuit Problems

Explore together: how do circuits contribute to urban planning, mail routes, and bus routes? Can your new knowledge allow you to solve the Chinese Postman Problem? How do Euler's circuits compare to Hamiltonian circuits? Can you solve the Traveling Salesman problem? If you're not familiar with these problems, then find here an opportunity to research and learn together!

Activity 7.2 Logic Puzzles

Pick out some logic problem from a library book or from online. Work together to discover relevant information and draw logical conclusions!

Activity 7.3 Business Profit/Loss

Determine a business you might like to own. Research the costs of buying a business. Explore a P&L or "Profit and Loss" sheets that might be available to investors seeking to purchase a business. Design a plan to buy and improve the business. What are the costs involved? What are the possible sources of money (grants, investors, loans, etc.)?

Activity 7.4 Voting Fairness

Economist Kenneth Arrow claims that an actually, truly fair voting system is not possible. What are the theories he explored? How did he base this assertion? Do you agree?

Activity 7.5 Payoff Matrices

What is a payoff matrix? What are some situations where you might use a payoff matrix? How would you fare in the classic Prisoner's Dilemma two-person variable-sum game? What is a zero-sum game?

Activity 7.6 How does GPS work?

Investigate the calculations, satellites, and the mathematics behind location finding using mobile technology.

Chapter 8

Preparing Late Teens

8.1 Development of 16-18-Year-Olds

In their later teen years, the focus on "black and white" situations begin to wane as teenagers begin to understand nuance and extenuating circumstances within scenarios. They are becoming more acutely aware of what those around them think. They are beginning to solve increasingly complex problems. There comes much trial and error in this period due to the prefrontal cortex not being fully developed. On the positive side, the self-involvement of the earlier years is beginning to subside as late teens begin to think more about others and more globally about how people and events interact to make the world.

8.2 Mathematics for 16-18-Year-Olds

At this time, some students diverge into different courses. While some students may move forward with an advanced algebra curriculum, others will take career-oriented courses such as computer programming, statistics for future scientists, or financial literacy. For those on a more traditional college preparatory track, students will deepen their understanding of the algebra, geometry, and trigonometry introduced in the last two years.

Polynomials and equations and functions are further developed and understood now. Series and sequences are explored as well as more advanced ideas from probability and statistics. It is also in this time that logarithms are introduced and sued to calculate exponents.

8.3 Sample Problems for 16-18-Year-Olds

Problem 8.1

What is the relationship between the sides of an isosceles right triangle and its hypotenuse?

Possible Solutions

Strategy: Look for a Pattern

In this strategy, choose a number of possibilities for the lengths of the legs of a triangle such as 3, 17, 21.5, 39, 526, and 9999. Use the Pythagorean Theorem to determine the length of the hypotenuse in each case.

Leg 1	Leg 2	Hypotenuse
3	3	$3\sqrt{2}$
17	17	$17\sqrt{2}$
21.5	21.5	$21.5\sqrt{2}$
39	39	$39\sqrt{2}$
526	526	$526\sqrt{2}$
9999	9999	$9999\sqrt{2}$

It seems that each example chosen, the legs are multiplied by the square root of 2 in the hypotenuse column.

Strategy: Convert to Algebra

If you let a be the length of the sides of the isosceles triangle and h be the hypotenuse, applying the Pythagorean theorem will produce the following algebraic representation:

$$a^2 + a^2 = h^2$$

Combining like terms returns:

$$2a^2 = h^2$$

Taking the square root of both sides finds:

$$\sqrt{2}a = h$$

At this point, you might say that the hypotenuse is always radical 2 times the length of a side in an isosceles triangle.

Problem 8.2

A teacher is predicting that 3 out of her class of 30 will score over 95 on an examination. If the mean of the exams was 86 and the standard deviation was 4, was her prediction accurate?

Possible Solutions

Strategy: Organize Information

This question lends its support on understanding the normal curve. The importance of the standard deviation is that it tells us how many results fall within each percentage of the mean.

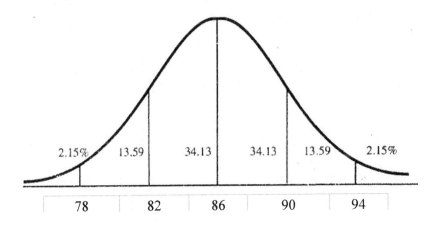

This depiction of the bell curve demonstrates that only 2.15% of students scored higher than a 94 on the test according to the data provided. That calculates to less than one student. The teacher's prediction was not accurate.

Problem 8.3

Use your knowledge of patterns to answer the following:

- What is the last letter in this pattern: **A Z C X E V G?**
- What are the next three numbers in this pattern: 2 7 4 9 6 11 8?
- What is the missing number: 120 ____ 109 102 94?

Possible Solutions

Strategy: *Look for a Pattern*

In the first pattern the letters of the alphabet are being systematically disbursed. The fact that the first two letters are A and Z can give us a clue, and the third letter being C means we skipped B and that pattern continues with E as we skipped D. As the letters are skipping forward, so are the letters counting down from Z. The pattern will continue:

A Z C X E V G T I Q K O M

For the second bulleted pattern, note that every other number is created by adding two to the number before its predecessor. We see that 2, 4, 6, 8 is interchanged with 7, 9, 11. So, the next three numbers will continue as 13, 10, and 15.

And for the final pattern, we find that the differences are increasing. If we start with 109, we must subtract 7 for a result of 102. Next we must subtract 8 to get 102. The missing number is 115 because it is five less than 115 and six more than 109.

Problem 8.4

You discover your clock is running fast, adding 6 minutes to each hour. The clock is set to 2:00 PM. What will the actual time be when your clock shows 8:00PM? What time would be the time on the clock when the actual time is 1:00 AM? *Challenge*: Will the clock ever show the correct time? If so, when? If not, explain why.

Strategy: *Create a Physical Representation*

There might be a clock that you could use to display this data. You might possibly have a toy clock or a non-working clock, or one that you might remove the battery to use. Otherwise, you could create a sketch and use household items to follow the broken clock:

Possible Solutions

Strategy: Create a systematic list

In this strategy, we will compare the fast clock's time to the real time each hour until our target of 8:00 PM

Fast Clock Time	Actual Time
2:00 PM	2:00 PM
3:00 PM	2:54 PM
4:00 PM	3:48 PM
5:00 PM	4:42 PM
6:00 PM	5:36 PM
7:00 PM	6:30 PM
8:00 PM	7:24 PM

Continuing in this pattern, by 1:00 AM real time, the clock will read 11:54 PM.

Problem 8.5

Speedy Phone is charging $30 for monthly mobile phone service plus $0.02 per MB of data used. Phones Plus is charging $60 plus $0.01 per MB of data used. Both companies have equal overage charges per MB. Under what circumstances would you choose each company? What amount of increase in overage charges would make your decision change?

Possible Solutions

Strategy: Guess and Check

If this was truly a situation you encountered in real life, this problem might be best solved Guess and Check. In a personal circumstance, you might look over your phone plan and see how much data used, on average.

Strategy: Convert to Algebra

These two situations can be compared by setting the following two equations equal to each other:

$$\text{cost} = 30 + 0.02\text{mb}$$
$$\text{cost} = 60 + 0.01\text{mb}$$

Strategy: Create a Model

Representing this situation on a graph can be helpful to visualize the point where the plans are the same cost and the amount of data that would be a better value with each plan. The following graph depicts the following linear functions:

$$\text{cost} = 30 + 0.02\text{mb}$$
$$\text{cost} = 60 + 0.01\text{mb}$$

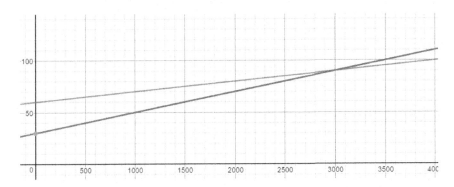

8.4 Everyday Problem Solving 16-18-Year-Olds

At this age, mathematics should relate as much as possible to real-life. Applications to finances and science are truly helpful for teens to explore how mathematics is used in their world.

Activity 8.1 Stock Market Game

Begin with an equal amount of money. Read up on current trends and news in the stock market. Choose how you will invest your investment. Watch both the value of the chosen stocks as well as the overall market analysis and investigate news. Did particular resources lead to make positive impact on earnings? Which sources were most reliable?

Activity 8.2 Investing Sense Explore

What happens to your money when you invest in different ways. What are the types of investing? How do bonds and stocks work? What are the benefits of a Money Market account?

Activity 8.3 Fractals

Explore the work of Benoit Mandelbrot and his work with fractals. Create your own using shareware, paper, or online graphing software.

Activity 8.4 Taxing Mathematics

Choose a career or business. Research typical salaries and expenses. Complete a basic or business tax form together. Modify the numbers to see how having children, a spouse, a home, etc. impact the result.

Activity 8.5 Another Line Way?

Most of us explored Euclidean geometry. Two mathematicians have proposed that parallel lines just may meet at some point. Explore how the impact on geometry when Euclid's fifth postulate is replaced by Riemann's or Bolyai & Lobachevsky's postulate.

Activity 8.6 Logical Thinking

Practice the laws of logic when discussing points with your child. Discuss the use of converse statements, straw man fallacy, contrapositive statements, generalizations, inferences, logical equivalences, inverse statements, contrapositive, conditional, biconditional, red herring, etc.

Activity 8.7 Income Taxes

Investigate tax codes and regulations. How does being married, single, or having children impact your taxes? When should you file different forms? What is the percentage of tax for each income bracket? What is deductible? What is an income statement? How are taxes different for self-employed, contractors, and business owners? What is a Schedule A and B?

Activity 8.8 Lotto Probability

Investigate the probabilities of winning different levels of lottery prizes. Consider multi-state lotteries. Review state lotteries as well as scratch-off lotteries. What makes a lottery prize more likely? How does the cost of the ticket relate to the probability of winning? Calculate the expected value or expected payoff for different games. *Extension*: Add casino games to your research. What is the difference between the probabilities of winning at slot machines versus card games versus roulette? How does your knowledge of these calculations impact your impression of casinos and lottery?

Chapter 9

Post-Secondary Problem Solvers

9.1 Development After Age 18

Research has demonstrated that significant changes (particularly to areas of the brain known for integrating cognition and emotion) are made in the brains of college students. While many neuroscientists agree that the brain is fully developed at the age of 25, others say there is continued growth until the mid-thirties. While some may not realize this, it is becoming more widely accepted that by graduation from high school, humans are only halfway through the process of maturity.

You may have heard of the prefrontal cortex. The executive functions of goal-setting, self-reflection, emotional regulation, and foresight are carried out here. This region is being developed through the middle of one's twenties. The study of the brain has revealed that the nerve fibers are covered with myelin. This substance helps transmit signals in the brain by acting as insulation. This process is one aspect of the physical nature of brain development. As this is happening, connections in our brains are also being strengthened, allowing us to communicate more efficiently.

9.2 Mathematics for 19-23-Year-Olds

There are many paths that one might take after compulsory education or graduation. Tradespeople become skilled in the mathematics necessary to build and to create. Sports managers might become more interested in how statistics impact their game and take up advanced learning. Elementary educators will learn more about how mathematics is developed in children and ways to make this mathematics accessible and relevant. While there is no one path that an adult might take, there are many mathematical ideas that will permeate the existence of citizens contributing to society:

- Budgeting
- Data analysis
- Discounts and markups
- Economics
- Investing
- Loan amortization
- Planning
- Probability
- Return on Investment
- Samples and statistics
- Stock market
- Taxes

9.3 Sample Problems for 19-23-Year-Olds

These are some problems for this age group:

Problem 9.1

On a certain day last week I had a doctor appointment, dropped off my dry cleaning, checked out a museum exhibit, picked up produce from the local farmer's market, and ate brunch at Joe's Place Restaurant. I know that my dry cleaning would be ready in exactly one week. I am trying to remember which day of the week I can pick up my dry cleaning. The doctor's office is closed on weekends and Joe's is closed on Mondays. The farmer's market is only open on Mondays, Wednesdays, and Fridays. The museum exhibit was open to the public Tuesdays, Fridays, and Sundays. What day should I pick up my dry cleaning?

Strategy: Eliminate Possibilities

Draw a chart of the seven days:

Sunday	Monday	Tuesday	Wednesday	Thursday	Friday	Saturday

For each day, annotate the chart with the days that are not possible based on the given information.

Strategy: Make a Systematic List

Create a list of each event, and annotate the event with the days of the week that are possible.

Errand	Possible days
Doctor appointment	
Dry cleaner's	
Museum exhibit	
Farmer's market	
Brunch at Joe's	

Strategy: Organize Information

Create a chart depicting both the errand and the day of the week. Complete the chart based on the given information.

	Sun	Mon	Tues	Wed	Thur	Fri	Sat
Doctor appointment							
Dry cleaner's							
Museum exhibit							
Farmer's market							
Brunch at Joe's							

Possible Solutions

The doctor appointment could not be on a weekend so Sunday and Saturday are marked as not possible in the doctor appointment row. Joe's is closed on Mondays so Monday is marked with an "X" in the Monday column. Since the Farmer's Market is only open three days, the other days, Sunday, Tuesday, Thursday, and Saturday must be marked as not possible. The same procedure is necessary for the Museum, which must be marked as not possible on Monday, Wednesday, Thursday, and Saturday.

	Sun.	Mon.	Tues.	Wed.	Thurs.	Fri.	Sat.
Doctor appointment	X						X
Dry cleaner's							
Museum exhibit		X		X	X		X
Farmer's market	X		X		X		X
Brunch at Joe's		X					

The only day all of the locations was open was Friday. Therefore, the dry cleaning will need to be picked up on Friday.

Problem 9.2

Three mothers have a total of 15 children including 9 boys. Kelly has three girls and Jenna has the same number of boys. Jenna has one more child than Kelly, who has four children. Nancy has four more boys than girls and the same number of girls as Kelly has boys. How many boys do Nancy and Kelly have?

Strategy: Organize Information

If there are three mothers, three lists can be made using the given information. Each of the clues found in the problem can be applied mathematically based on each mother in order to determine the number of children for each mom.

Strategy: Make a Systematic List

The given information starts with the fact that there are three parents with fifteen children and nine boys in the group. The following chart indicates this initial information. Complete the chart using the remaining given information.

	Girls	Boys	Total
Kelly			
Jenna			
Nancy			
Total		9	15

Possible solution

In the first step, we can calculate the number of girls by subtracting the number of boys, 9, from the total children, 15.

	Girls	Boys	Total
Kelly			
Jenna			
Nancy			
Total	6	9	15

In the next step we can fill-in that Kelly has three girls and Jenna has 3 boys from the information in the second sentence.

	Girls	Boys	Total
Kelly	3		
Jenna		3	
Nancy			
Total	6	9	15

In the next step we can fill in that Kelly has four children and Jenna has five children from the information given in the third sentence.

	Girls	Boys	Total
Kelly	3		4
Jenna		3	5
Nancy			
Total	6	9	15

In the next step, the fourth sentence provides us information about Nancy. The information is not numerical but can be used to solve the remaining missing numbers.

	Girls	Boys	Total
Kelly	3		4
Jenna		3	5
Nancy	Same as Kelly's boys	Girls + 4	
Total	6	9	15

From our chart, we can calculate that Nancy has 6 children because Jenna and Kelly have 9 combined. We can also calculate that Kelly has 1 boy because she has 4 children and three are girls. Finally, we know that Jenna has 2 girls because she has five children and 3 are boys.

	Girls	Boys	Total
Kelly	3	1	4
Jenna	2	3	5
Nancy	Same as Kelly's boys	Girls + 4	6
Total	6	9	15

This leaves Nancy's children. We can calculate using the columns or the given information. Either way, Nancy has 1 boy (6-3-2 or same as Kelly's) and she has 5 boys (her girls plus four or 9-1-3).

	Girls	Boys	Total
Kelly	3	1	4
Jenna	2	3	5
Nancy	1	5	6
Total	6	9	15

We have determined everyone's children and their sexes however we have not yet answered the question! The question asks "How many boys do Nancy and Kelly have?" The answer to that question is 6.

Problem 9.3

A daughter gave her parents a $40 gift card for a local diner. The three family members went out and each ate a special, but the daughter does not eat red meat. The specials menu were: BBQ Shrimp: $18, Chicken Florentine: $14, Cheese Steak: $12. What possible orders could the family have ordered without exceeding the balance on the gift card?

Strategy: Make a Systematic List

Create a list indicating who orders which meal. Be certain that no one orders a meal that will put the order over the indicated amount of the gift card.

	Number of Orders				
Shrimp					
Chicken					
Steak					
Cost					

Strategy: Identify Sub-Problems

For this strategy, determine the meals or combinations that would be problematic. That is, which meal combinations cannot be allowed? Avoiding certain meal combinations would allow for precision in choosing. For example, clearly all three family members cannot choose the shrimp dish because the bill would be $54. All three family members cannot choose the chicken dish because the bill would be $42. Two shrimps and a chicken must be eliminated because the total would be $50. In fact, any order that includes the shrimp dish will become over budget.

Strategy: Use Matrix Logic

Matrices can be useful when determining the cost spread of multiple calculations.

$$\begin{bmatrix} 18 \\ 14 \\ 12 \end{bmatrix} x \begin{bmatrix} 3 & 0 & 0 & 2 & 2 & 1 & 1 & 0 & 0 & 1 \\ 0 & 3 & 0 & 1 & 0 & 0 & 2 & 1 & 2 & 1 \\ 0 & 0 & 3 & 0 & 1 & 2 & 0 & 2 & 1 & 1 \end{bmatrix} =$$

$$\begin{bmatrix} 54 & 0 & 0 & 36 & 36 & 18 & 18 & 0 & 0 & 18 \\ 0 & 42 & 0 & 14 & 0 & 0 & 28 & 14 & 28 & 14 \\ 0 & 0 & 36 & 0 & 12 & 24 & 0 & 24 & 12 & 12 \end{bmatrix}$$

Taking the sum of each column would determine the total cost of each set of choices.

Strategy: Act It Out

Have three people take on the roles of parents and child. Take turns choosing meals, ensuring the total never exceeds $40. Try multiple rounds to see how many different ways the family can order. If the order is too expensive for the third person to make a choice, eliminate that choice and try again.

Possible Solutions

There are a number of possible choices that three people might make at this restaurant.

	Number of Orders									
Shrimp	3	0	0	2	2	1	1	0	0	1
Chicken	0	3	0	1	0	0	2	2	1	1
Steak	0	0	3	0	1	2	0	1	2	1
Cost	$54	$42	$36	$50	$48	$42	$46	$40	$38	$44

There are only three possible combinations of orders that stay within the $40 budget: three steak meals, two chicken and one steak, and one chicken and two steak.

Problem 9.4

There are a group of trainers at a pet training facility. Jem trains poodles and German shepherds, Sandra trains Pugs and Black Labs, Kem trains Collies and Poodles, Gia trains German shepherds and pugs. If Poodles are easier to train than German shepherds, Black Labs are harder than pugs, German shepherds are easier than pugs, and Collies are easier than Poodles, then which trainer train the most difficult dogs?

Strategy: Eliminate Possibilities

Each time the problem indicates that a dog 'is easier than,' the dog on the easier end is eliminated. Continue with all of the clues until only one remains.

Strategy: Organize Information

This might be done by putting the name of each dog and its' trainer on index cards. The index cards can be arranged from easiest to most difficult, and can be moved without erasing or writing over already-written information.

Possible Solutions

Step 1: Poodles are easier than German Shepherds

Easier
- Poodles (Jem & Kem)
- German Shepherds (Gia & Jem)

Harder

Step 2: Black Labs are harder than pugs

Easier
- Pugs (Gia & Sandra)
- Black Labs (Sandra)

Harder

Step 3: German shepherds are easier than pugs.

Easier
- Poodles (Jem & Kem)
- German Shepherds (Gia & Jem)
- Pugs (Gia & Sandra)
- Black Labs (Sandra)

Harder

Step 4: Collies are easier than Poodles.

Easier
- Collies (Kem)
- Poodles (Jem & Kem)
- German Shepherds (Gia & Jem)
- Pugs (Gia & Sandra)
- Black Labs (Sandra)

Harder

Step 5: Since Sandra trains the most difficult type of dogs, she would be considered the trainer to be the solution to this problem.

Problem 9.5

Your supervisor would like to target a postcard ad for a new product to a particular audience: licensed hair stylists who also drive. You have researched that there are 27,000 people in the town. There are 49 stylists that do not drive and 20,299 drivers who are not stylists. 20,471 people in the town are drivers but not stylists. Finally, 6,210 are neither stylists nor licensed. How many postcards should you print?

Strategy: Draw Venn Diagram

For this problem, you are put into the place of a marketing employee. There are multiple steps to this problem and a Venn diagram will help to organize the information:

The inside of the box represents the number of people in the town. The circle on the left represents the licensed drivers, the circle on the right represents the stylists. The overlap represents licensed, driving stylists.

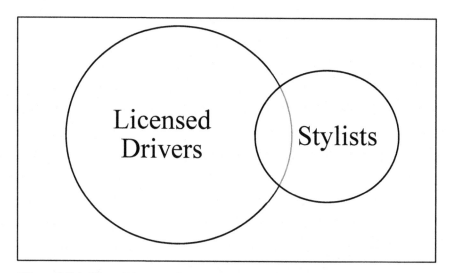

Figure 9.3-1: Venn Diagram for Problem 9.5

Possible Solutions

The given information tells us that there are 27,000 people in the town and there are several subtractions that should be done in order to find the number of stylists who drive. From the total 27,000, subtract the 49 stylists that don't drive, the 20299 drivers who are not stylists, and the 6210 people who neither drive or style. The remaining number is 442, however since this number applies to both circles, we must divide it in half. You should print 221 postcards to ensure that every driving stylist in the town gets the ad.

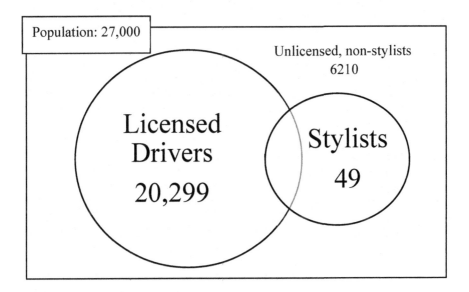

9.4 Everyday Problem Solving 19-23-Year-Olds

Activity 9.1 Time Management

People sometimes claim they do not have time for certain activities or events. There are 168 hours in each week. Make a chart of what you do each hour for one week. Create a pie chart to depict how you spend your time over the course of one week. Categorize actions like sleep (to include naps), eating (all meals/snacks), work (including overtime or working from home), commuting, etc. Analyze how you are spending your time. Would you prefer a different balance? Is there something you'd like to add in order to become a better you (meditating, reading, taking classes, fitness, spending time with loved ones, etc.)? Create a plan to replace the activities you wish to reduce.

Activity 9.2 Purchase Planning

Analyze a rather large future purchase. Choose something you are considering investing in (a home, vehicle, degree, travel, etc.) Investigate the costs or the amortization schedules of loans for the purchase (which ever apply). How long would it take for you to save for the down payment? What is the cost of the item's interest? Could you save in advance to avoid paying interest?

Activity 9.3 Mindful Living

Make a list of your top ten priorities. Analyze your daily, weekly, monthly, and yearly spending. Are you spending your money on what you purport to prioritize?

Activity 9.4 Entrepreneurship Investigation

Choose a business or franchise you would like to open or purchase. Research local businesses for sale. Review the Profit & Loss sheets and investigate making a wise investment. Create a business plan to present to a potential financer.

Activity 9.5 Keeping Your Money

Investigate options from local banks for keeping your money. Review the benefits and costs for the different types of accounts and options such as bonds, money market accounts, checking accounts, and savings accounts. Prepare a presentation for a friend of future spouse on where you would keep your monies in different increments (for example: what if you had $5,000 to store versus $50,000 or $500,000). Understand what the FDIC is and how it impacts your decisions.

Activity 9.6 Planning for Retirement

Research the options for investing your money in preparation for retirement. What are the possible benefits of different types of jobs? How does a 403b differ from a 401k? How does a Roth IRA differ from an IRA? How do you accrue Social Security? How and when do you collect Social Security? Research then create a goal for retiring and a long-term plan to meet the goal.

Activity 9.7 Automobile Insurance

Choose a make and model of a car. Determine how you will choose your coverages and deductibles. Then find out: What does insurance cost for this car from different companies? How does your age, gender, and history impact the cost? How can you get discounts on your insurance (through employers, courses, course grades, etc.)?

Activity 9.8 Insurance

While automobile insurance is a requirement, there are many other types of insurances to investigate. There's renter's insurance and personal property insurance. There are umbrella policies. There are whole-life and term life insurance policies. There are supplemental and long-term care insurance policies. What types of insurance are necessary? What types are mandatory? What types would you like to plan to have?

Activity 9.9 Large Purchase Planning

Consider something you might like to do in the future: purchase and own a boat, restore a classic car, build a pool and outdoor kitchen, redesign your own brand new bathroom, travel to three continents for six months, earn another degree, design, patent, and sell a new invention. Before doing any research, write down what you predict this larger purchase will cost and how long it would take to implement. Then research the components and costs of the project. Read blogs or articles from folks who have done what you'd like to do. Reflect on whether the cost and the time is something you could prepare for and how long it will take you to do so.

Chapter 10

Conclusion

The adults in a child's life have a tremendous opportunity to foster problem solving skills that are relevant in all areas of productive living while also allowing students to develop mathematical thinking. The authors of the Common Core State Standards developed eight Standards for Mathematical Practice that are expected to be developed across curricular topics and in all grade levels:

- Make sense of problems and persevere in solving them.
- Reason abstractly and quantitatively.
- Construct viable arguments and critique the reasoning of others.
- Model with mathematics.
- Use appropriate tools strategically.
- Attend to precision.
- Look for and make use of structure.

These practices can lead to a reduction in mathematics anxiety since they emphasize the process of doing mathematics over answer-getting. Developing these practices will lead to increased understanding of mathematics while addressing both mathematical problem solving and problem solving as a life skill.

There is a level of necessity in being resilient and finding solutions to difficult problems in life and in mathematics. There is also a level of confidence that can be life-changing when young people find themselves achieving in mathematics, particularly with word problems.

Glossary

>	sign of inequality known as the greater-than sign. The quantity on the left of the sign is of greater quantity than the one on the right
<	sign of inequality known as the less-than sign. The quantity on the left of the sign is of smaller quantity than the one on the right
≥	sign of inequality known as the greater-than-or-equal-to sign. The quantity on the left of the sign is of greater quantity or equal to the one on the right
≤	sign of inequality known as the less-than-or-equal-to sign. The quantity on the left of the sign is of smaller quantity or equal to the one on the right
Absolute value	the magnitude of a real number; distance a real number from zero on the number line. Symbol: $|x|$

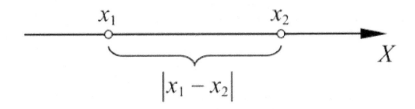

Figure G-1: Absolute Value

Add	to put together to calculate a total value
Addend	a number that is added to a number
Adjacent angles	two angles are adjacent when they do not overlap and have a common side and a common vertex

Figure G-2: Adjacent Angles

Angle	a shape, formed by two lines or rays diverging from a common point, called the vertex
Area	the measure of how much space there is on a flat surface
Area model	model for mathematics used to determine the area where the length and width represent the multiplicand and multiplier
Array	a systematic arrangement of similar objects, usually in rows and columns.
Associative Property	property that holds for addition and multiplication, indicating that you can add or multiply regardless of how the numbers are grouped

$$(x + y) + z$$
$$=$$
$$x + (y + z)$$

Figure G-3: Associative Property of Addition

Bivariate data   data that has two variables
Blocks   solid piece of hard material with flat surfaces on
                      each side

Figure G-4: Blocks

Brackets   mathematical notation such as parentheses ( ),
                      square brackets [ ], braces { }, and angle brackets
                      ⟨ ⟩ indicating grouping of the expression within
Cardinal   a number denoting quantity (one, two, three, etc.)
Categorical data   data that can be categorized or groups (as
                      opposed to numerical data). Examples might be:
                      Eye color, School, Town, etc.
Circle   simple closed shape in Euclidean geometry that is
                      the set of all points that are of a given distance
                      (radius) from a given point (center)
Circumference   the enclosing boundary of a curved geometric
                      figure, especially a circle
Coefficient   a numerical or constant quantity placed before
                      and multiplying the variable in an algebraic
                      expression

Commutative Property   property that holds for multiplication and addition that indicates changing the order of the operands does not change the result

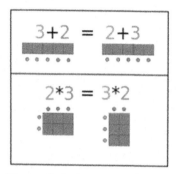

Figure G-5: Commutative Property

Congruent   two figures are congruent if they are equal in size and shape

Construction   the process of drawing shapes, angles, or lines accurately

Coordinate plane   two-dimensional surface with two scales, a horizontal scale called the x-axis and vertical scale called the y-axis

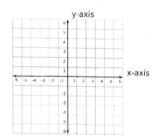

Figure G-6: Coordinate Plane

Counters   items or objects used for counting

Cube   three-dimensional object bounded by six equally sized square faces

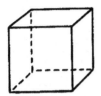

Figure G-7: Cube

Decimal	a number denoting a system of numbers and arithmetic based on the number ten, tenth parts, and powers of ten
Dependent variable	a variable (often denoted by y) whose value depends on that of another.
Diagram	representation of information in graphic form
Digit	any of the numerals from 0 to 9, especially when forming part of a number
Dilation	a transformation that produces an image that is the same shape as the original, but is a different size, multiplied by a scale factor
Distributive Property	states that multiplying a sum by a number is the same as multiplying each addend by the number and then adding the products: $a \times (b + c) = (a \times b) + (a \times c)$
Divide	separate or be separated into parts
Dot plot	a statistical chart consisting of data points plotted on a scale, typically using dots or Xs

Dot plot of Random Values

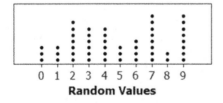

Equation	a statement that the values of two mathematical expressions are equal (indicated by the sign =)
Equivalent fractions	different fractions that represent the same quantity e.g., ½ and $^2/_4$ and $^{16}/_{32}$ are all equivalent
Equivalent Ratio	different ratios that represent the same ratio e.g., 2:7 is an equivalent ratio to 20:70 as well as 6:21
Evaluate	find a numerical expression or an equivalent by performing operations
Exponent	how many times the number will be multiplied.
Expression	Numbers, symbols and operators (such as + and ×) grouped together
Exterior angle	the angle between one side of a triangle and the extension of an adjacent side
Factor	a number or quantity that when multiplied with another produces a given number or expression.

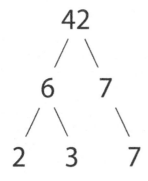

Figure G-8: Factors of 42

Fraction	part of a whole
Frequency	the rate at which something occurs or is repeated over a particular period of time or in a given sample
Function	a relation from a set of inputs to a set of possible outputs where each input is related to exactly one output

Greatest common factor	the greatest whole number that divides two numbers evenly
Independent variable	a variable (often denoted by x) whose variation does not depend on that of another
Inequality	a formal statement of inequality between two quantities usually separated by a sign of inequality (as $<$, $>$, or $\neq$ signifying respectively is less than, is greater than, or is not equal to)
Inference	conclusion or opinion that is formed because of known facts or evidence
Interquartile range	the difference between the 3rd and 1st quartiles in a data set, each of which is the median of the upper and lower halves of the data, respectively
Inverse	operation that reverses the effect of another operation. E.g., addition and subtraction are inverse operations as are multiplication and division
Irrational number	a real number that cannot be expressed as a ratio of integers e.g., $\sqrt{7}, \pi, \sqrt[4]{2}$
Line segment	part of a line that is bounded by two distinct end points, and contains every point on the line between its endpoints
Mass	measure of how much matter is in an object
Mean	is the average of the numbers calculated by adding up all the numbers, then dividing by how many numbers there are
Mean absolute deviation	the average distance between each data value and the mean
Median	the middle number in a data set with an odd number of elements or the average of the two middle numbers in a data set with an even number of elements

Mode	the most frequently occurring number in a data set
Model	a description of a system using mathematical concepts and language
Modeling	the process of developing a mathematical model
Multiple	a number multiplied by an integer
Multiply	the process of calculating the result when a number a is taken b times
Negative numbers	a real number that is less than zero
Net (of three-dimensional figures)	a pattern that can be cut and fold to make a model of a solid shape
Number bonds	a mental picture of the relationship between a number and the parts that combine to make it
Number cube	another name for die or dice
Number Line	a line on which numbers are marked at intervals, used to illustrate simple numerical operations
Order of operations	a collection of rules that define which procedures to perform first in order to evaluate a given mathematical expression
Ordinal	a number denoting order (first, second, third, etc)
Outcome	a possible result of a probability experiment
Parallel	two lines that are always the same distance apart and never touch
Pattern blocks	a type of mathematical manipulatives that allow children to see how shapes can be decomposed into other shapes and introduces them to tiling

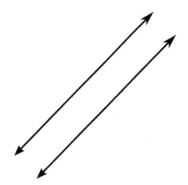

Figure G-9: Parallel Lines

Figure G-10: Pattern Blocks

Percent increase and decrease — measures of percent change, which is the extent to which a variable gains or loses intensity, magnitude, extent, or value and is calculated by dividing the difference in the two measures by the original value

Percent — "per" means for each and "cent" means 100, so this is a ratio whose second term is 100 and means "per hundred"

Perimeter	the total distance around the edge of a figure
Perpendicular	means to be at right angles
Picture graph	a pictorial display of data with symbols, icons, and pictures to represent different quantities
Place value	the value of the place, or position, of a digit in a number
Polygon	any closed 2-dimensional shape formed with straight lines
Prism	A solid object with two identical ends and flat sides
Probability	the likelihood of something happening in the future
Product	the answer of an equation in which two or more variables are multiplied
Proportion	a statement that two ratios are equal
Pythagorean Theorem	a fundamental relation in Euclidean geometry among the three sides of a right triangle which states that the square of the hypotenuse (the side opposite the right angle) is equal to the sum of the squares of the other two sides
Quadrant	any of the 4 areas made when we divide up a plane by an x and y axis (as shown), named using Roman Numerals

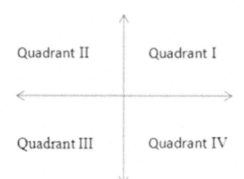

Figure G-11: Quadrant

Quadrilateral	a four-sided polygon with four angles
Quartile	the values that divide a list of numbers into quarters
Quotient	the result obtained by dividing one quantity by another
Radical	a mathematical expression indicating a root
Ratio	a comparison of two numbers by division
Rational number	any number that can be expressed as the quotient or fraction p/q of two integers, a numerator p and a non-zero denominator q
Ray	a line which starts at a point with given coordinates, and goes off in a particular direction to infinity
Rectangle	a 4-sided flat shape with straight sides where all interior angles are right angles and whose opposite sides are parallel and of equal length
Rectangular prism	a solid (3-dimensional) object which has six faces that are rectangles

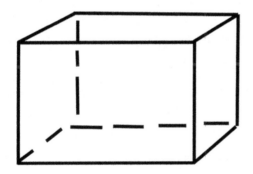

Figure G-12: Rectangular Prism

Reflection	a transformation in which a geometric figure is reflected across a line, creating a mirror image. That line is called the axis of reflection

Teaching Children to Love Problem Solving

Rhombus   a 4-sided flat shape with straight sides where all sides have equal length and whose opposite sides are parallel and opposite angles are equal

Rotation   a transformation in which a plane figure turns around a fixed center point

Ruler   an instrument used in geometry, technical drawing, printing, engineering and building to measure distances or to rule straight lines

Sample population   a selection taken from a larger group (the "population") that can be used to examine it to find out something about the larger group

Sample space   the set of all possible outcomes or results of an experiment

Scatter plot   a graph of plotted points that show the relationship between two sets of data

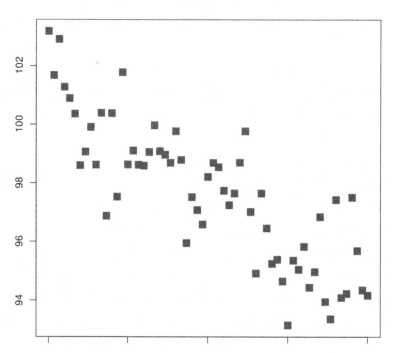

Figure G-13: Scatter Plot

Scientific notation	a method of writing or displaying numbers in terms of a decimal number between 1 and 10, multiplied by a power of 10
Sequential	of, relating to, or arranged in a particular order or sequence. : happening in a series or sequence
Set	a well defined collection of distinct objects, considered as an object in its own right
Simple interest	the interest calculated only on the principal regardless of the interest earned so far
Simulation	a way to model random events, such that simulated outcomes closely match real-world outcomes
Skip count	Counting forwards or backwards by a number other than 1
Slope	a measure of the steepness of a line, or a section of a line, connecting two points
Solid	three dimensional object; dimensions are called width, depth and height. Examples include spheres, cubes, pyramids and cylinders
Sphere	the set of all points equidistant from a single point in space; a geometric figure that is perfectly round
Square units	units to measure area
Straightedge	a tool with an edge free from curves, or straight, used for transcribing straight lines, or checking the straightness of lines
Subtract	operation that represents the operation of removing objects from a collection. It is signified by the minus sign ($-$)
Supplementary angles	angles whose measures sum to 180 degrees
Surface Area	the sum of the areas of the faces of a three dimensional figures
Symmetry	one shape becomes exactly like another when you move it in some way: turn, flip or slide
Tally (tallies)	numeral marks used for counting

Tape diagram	visual models that use rectangles to represent the parts of a ratio
Term	a single number or variable, or numbers and variables multiplied together. Terms are separated by + or − signs
Tessellate	a flat surface is the tiling of a plane using one or more geometric shapes, called tiles, with no overlaps and no gaps
Translation	a function that moves an object a certain distance. The object is not altered in any other way. It is not rotated, reflected or re-sized
Transversal	a line that passes through two lines in the same plane at two distinct points
Trapezoid	four-sided flat shape with straight sides that has a pair of opposite sides parallel
Triangle	a polygon with three edges and three vertices
Unit	type of measurement
Unit fraction	a rational number written as a fraction where the numerator is one and the denominator is a positive integer
Unit Rate	a rate with 1 in the denominator. If a rate is given and the quantity in the denominator is not 1, unit rate is calculated by finding an equal ratio such that the denominator is 1
Unknown	a variable, or the quantity it represents
Variability	a measure of the spread of a data set
Vertex	a point where two or more straight lines meet

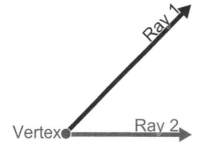

Figure G-14: Vertex

Vertical angles	each of the pairs of opposite angles made by two intersecting lines
Vertices	plural of vertex; points where two or more straight lines meet
Visualization	a technique for creating images or diagrams
Volume	the amount of space that a substance or object occupies, or that is enclosed within a container
y-intercept	the point on the coordinate plane where a relation intersects with the y-axis
Zero	whole number between -1 and 1 with the symbol 0 and indicating no amount

Bibliography

Allsopp, D. H., Kyger, M. M., & Lovin, L. H. (2007). *Teaching Mathematics Meaningfully: Solutions for Reaching Struggling Learners.* Brookes Publishing Company. PO Box 10624, Baltimore, MD 21285.

American Academy of Pediatrics. (2001). Media violence. *Pediatrics, 108*(5), 1222-1226.

Bahr, D. L., & DeGarcia, L. A. (2008). *Elementary Mathematics is Anything but Elementary: Content and Methods from a Developmental Perspective.* Cengage Learning.

Blitzer, R., & White, J. (2005). *Thinking Mathematically.* Pearson Prentice Hall.

Boaler, J. (2008). *What's Mathematics Got to Do with it: Helping Children Learn to Love Their Least Favorite Subject--and why It's Important for America.* Viking.

Brumbaugh, D. K., & Rock, D. (2010). *Teaching Secondary Mathematics.* Routledge.

Burns, M. (2000). *About Teaching Mathematics: A K-8 resource.* Mathematics Solutions Publications, Marilyn Burns Education Associates, 150 Gate 5 Road, Suite 101, Sausalito, CA 94965.

Bus, A. G., Van Ijzendoorn, M. H., & Pellegrini, A. D. (1995). Joint book reading makes for success in learning to read: A meta-analysis on intergenerational transmission of literacy. *Review of Educational Research, 65*(1), 1-21.

Cathcart, W. G. (2001). *Learning Mathematics in Elementary and Middle Schools.* Prentice Hall.

Consortium for Mathematics, & Its Applications (US). (2009). *For All Practical Purposes: Mathematical Literacy in Today's World.* Macmillan.

Copley, J. V. (2000). *The Young Child and Mathematics.* Washington, DC: National Association for the Education of Young Children.

Costello, M. J. (1988). *The Greatest Puzzles of All Time.* Courier Corporation.

Feldman, R. S., & Garrison, M. (1993). *Understanding Psychology* (Vol. 10). New York, NY: McGraw-Hill.

Hoffman, B. L., Breyfogle, M. L., & Dressler, J. A. (2009). The Power of Incorrect Answers. *Mathematics Teaching in the Middle School, 15*(4), 232-238.

Johnson, K., Herr, T., & Kysh, J. (2004). *Crossing the River with Dogs: Problem Solving for College Students.* Springer Science & Business Media.

Kennedy, L., Tipps, S., & Johnson, A. (2007). *Guiding Children's Learning of Mathematics*. Cengage Learning.

Kodaira, K. (1996). *Mathematics 1: Japanese Grade 10* (No. 8). American Mathematical Soc.

Montessori, M., & Claremont, C. A. (1969). *The Absorbent Mind*. New York: Dell Pub. Co.

National Council of Teachers of Mathematics (Ed.). (2000). *Principles and Standards for School Mathematics* (Vol. 1).

O'Shea, T. (1993). *Measuring Up: Prototypes for Mathematics Assessment*.

Polya, G. (2014). *How to Solve it: A New Aspect of Mathematical Method*. Princeton University Press.

Posamentier, A. S., & Krulik, S. (2008). *Problem-solving Strategies for Efficient and Elegant Solutions, Grades 6-12: A Resource for the Mathematics Teacher*. Corwin press.

Posamentier, A. S., & Schulz, W. (1996). *The Art of Problem Solving: A Resource for the Mathematics Teacher*. Corwin Press, Inc., Thousand Oaks, CA.

Reys, R. E., Lindquist, M., Lindquist, M. M., Lambdin, D. V., & Smith, N. L. (2014). *Helping Children Learn Mathematics*. John Wiley & Sons.

Smith, S. S. (2012). *Early Childhood Mathematics*. Pearson Higher Ed.

Sowder, J. T., Sowder, L., & Nickerson, S. D. (2008). *Reconceptualizing Mathematics*. New York: WH Freeman.

Van de Walle, J. A., Karp, K. S., Lovin, L. A. H., & Bay-Williams, J. M. (2013). *Teaching Student-centered Mathematics: Developmentally Appropriate Instruction for Grades 3-5* (Vol. 2). Pearson Higher Ed.

Ward, R. A. (2009). *Literature-based Activities for Integrating Mathematics with Other Content Areas, Grades 6-8*. Pearson.

Whimbey, A., Lochhead, J., & Narode, R. (2013). *Problem Solving & Comprehension*. Routledge.

White, B. L. (1993). *The First Three Years of Life*. Touchstone.

Index